TAMING THE MUSKINGUM

TAMING THE MUSKINGUM

EMORY L. KEMP

WEST VIRGINIA UNIVERSITY PRESS
MORGANTOWN 2015

First edition published 2015 by West Virginia University Press
Printed in the United States of America

23 22 21 20 19 18 17 16 15   1 2 3 4 5 6 7 8 9

ISBN:
paper   978-1-940425-83-2
epub   978-1-940425-85-6
pdf   978-1-940425-84-9

Library of Congress Cataloging-in-Publication Data:
Kemp, Emory.
  Taming the Muskingum / Emory Kemp.
     pages cm
  ISBN 978-1-940425-83-2 (pbk.)—ISBN 978-1-940425-85-6 (epub)—ISBN 978-1-940425-84-9
(pdf)
  1. Muskingum River (Ohio)—Navigation. 2. Flood control—Ohio—Muskingum River—History. I.
Title.
  HE630.M6K47 2015
  386'.3097719—dc23
                                    2015011287

Book and cover design by Than Saffel
Cover image: The steamboat *Enterprise* is one of the earliest steamboats. (C. W. Sutphin collection)

The West Virginia University Press gratefully acknowledges the financial assistance provided for this book
by the Preservation Alliance of West Virginia, the statewide grassroots organization supporting historic
preservation in the Mountain State. To learn more about Preservation Alliance visit www.pawv.org.

CONTENTS

ACKNOWLEDGMENTS

For a number of years I have greatly benefited from support by the United States Army Corps of Engineers, Huntington and Pittsburgh Districts. This research and associated reports concentrated on the improvement of the principal Ohio River tributaries. I am especially grateful to the Pittsburgh District for support on the work published here. Other studies included *The Great Kanawha Navigation*, *The Little Kanawha Improvement*, *The Big Sandy Improvement*, *Building Tygart Dam*, as well as numerous smaller investigations, in particular the study by Charles Ellet Jr. regarding flood control by reservoirs and levees.

I need to declare the generous subvention for this book by the Preservation Alliance of West Virginia, the statewide organization involving historic preservation and education.

The Institute for the History of Technology and Industrial Archaeology was most supportive of all these studies listed above. For his involvement in all phases of this book it is dedicated to Larry N. Sypolt.

PREFACE

———

I n considering the history of attempts to control the Muskingum
Watershed, several approaches can be expected to yield new insights and
enhance appreciation for the subject.

Amongst the possibilities are a focus on the problems connected with
flood control; a case study of the role of the Corps of Engineers in New Deal
Public Works projects; the significance of the conservation movement with
regard to the Muskingum watershed; a study of the Muskingum Watershed
Conservation District as a pioneer in a comprehensive approach to water
resource management; and, for those interested in canals and waterways, the
Muskingum slackwater navigation, with particular emphasis on steamboats
and later gasoline-powered tows plying the river from Marietta to Dresden.
While touching on these subjects, this work emphasizes the engineering and
construction aspects of the Muskingum navigation and the fourteen related
flood-control dams built under the direction of the United States Army Corps
of Engineers. The beginning chapters focus on early settlement and attempts
to navigate the uncontrolled Muskingum River. The slackwater navigation,
which represented the first attempt to control the river for navigation purposes, occurred in conjunction with the building of the Ohio and Erie Canal
that stretched from Cleveland through the watershed area to Portsmouth on
the Ohio River.

Floods punctuate the story of the Corps of Engineers and its involvement
with flood control and navigation. With a well-thought-out plan for a system
of flood-control reservoirs, the Muskingum watershed qualified early in the
New Deal as a major public-works project under the direction of the U.S.
Army Corps of Engineers to provide unemployment relief. At the same time,
construction of the dams provided flood control. While this work does not

attempt to be an engineering text on the design and construction of earth and concrete dams, one chapter devoted to the subject presents the salient points of design and the state of the art in construction in the 1930s.

The text is richly illustrated with drawings, photographs, and maps showing many aspects of the dam and reservoir system as well as the Muskingum slackwater navigation. The appendix provides additional details on the various construction sites and the geology of the region.

A Wilderness Transformed

But the strongest call came from the canal, for the engineers were letting the water in for the first time today. As far as you could see, the big ditch ran, like a hill turned down and inside out. An army of Irishmen had scooped it out, cursing at the boys who threw stones at their clay pipes when they laid them up on the ground. Now they were gone and the ditch lay new and dry. But God help the dog or cat found in it when the water came down today. They were letting it in from the river. Oh, this would be a day to remember. Folks were coming from twenty miles to see boats floating where had been only dry earth before.

—The Town, Conrad Richter

At the end of the Ice Age, the great ice sheet covering much of Ohio receded as meltwaters draining toward the south carved out a dentrated drainage system. Through erosion over the millennia, this geological formation produced the landscape known to generations of Native Americans, who arrived in the region some 15,000 years ago.

Fur traders led European settlement as they penetrated the rich and heavily forested region, using the Ohio River and its many tributaries as the easiest means of transportation. Although not verified, the famous French explorer LaSalle may have explored the upper reaches of the Ohio Valley in 1669. Fur traders were increasingly active in the late seventeenth and early eighteenth centuries. Arnout Viele floated down the Allegheny and the upper reaches of the Ohio in 1692. Earlier in the century, from 1654 to 1664, Abraham Wood explored the tributaries of the Ohio. By the eighteenth century the French had penetrated the Ohio Valley utilizing the only practical means of transportation, rivers and lakes—the St. Lawrence, the Great Lakes, and the Ohio River itself. To counter the incursions of British traders and to reclaim the entire watershed of "La Belle Rivière" for the French crown, the Celoron expedition was launched in 1749. To secure the claim for the French, Celoron deposited metal plates at various locations along the Ohio River, one being buried at the confluence of Wheeling Creek and the Ohio River, and another at Marietta. It is not surprising that the French erected forts along the Allegheny River and elsewhere to link the Great Lakes with the Ohio Valley. In a countermeasure, Governor Dinwiddie, in an effort to defend claims by the colony of Virginia, built a fort at the confluence of the Allegheny and Monongahela Rivers, the present site of Pittsburgh, but the French drove out the Virginians and constructed Fort Duquesne.

This vying for territory in the Ohio Valley was a reflection of a larger struggle between Britain and France culminating in the Seven Years' War (a.k.a. the French and Indian War), 1754–1763.[1] With the Treaty of Paris in 1763, the French ceded all of French Canada and claims on the Ohio Valley. Three years after the treaty, an expedition left Pittsburgh with Captain Harry Gordon, chief engineer, and his assistant ensign Thomas Hutchin to map the river for use by the British military command. They were joined by George Morgan and George Croghan as commanders of the expeditionary fleet. The report of Gordon and Hutchin provided the first reliable information on the river. The opening of the American Revolution found Captain Hutchin in London preparing his report and maps for publication that eventually carried the date of 1788. Highly regarded not only by military personnel, but by many other settlers in the area, the maps and reports served for many years as the best information available on the Ohio River.

Following the American Revolution, the military engineer Colonel Rufus Putnam, former chief engineer of the Continental Army, was instrumental in organizing the Ohio Company in 1786. The purpose of the company was to settle demobilized soldiers on public lands. He is celebrated as the founder of Marietta, Ohio, at the mouth of the Muskingum River in 1788.[2] For added security Marietta, the first settlement in the entire Northwest Territory, was located across the Muskingum River from Fort Harmar, which was garrisoned by federal troops. The Ohio Company received 1.5 million acres that extended to the west and southward from the mouth of the Muskingum River. At about the same time as the founding of Marietta, Colonel Putnam began construction of a fortification called Campus Martius to provide protection for the town and to house members of the Ohio Company and their families.[3]

Despite his outstanding service to Britain, Hutchin fled London, where he was working on his report, for France. There he joined the American cause. He served as the geographer of the United States by appointment of the Continental Congress.[4] Following the Treaty of Paris, river traffic increased steadily on the Ohio and its tributaries despite the somewhat treacherous nature of a river strewn with boulders, snags, and shoals. Canoes, bateaux, and flatboats formed the motley array of vessels that made possible the early settlement of the Ohio Valley.[5] Following its founding in 1788, Marietta became an important center for boat building as a result of the availability of ship timbers. The Muskingum River, even at this early stage, saw keelboats bringing agricultural and forest products downriver during favorable river stages.

Under the Articles of Confederation, Congress established the Northwest Ordinances of 1785 and 1787, which established the means by which territories could become states with all the rights and privileges of each state's citizens. The ordinance forbade slavery, while establishing a method for surveying and selling lands lying north and west of the Ohio River. It was only a year later that Putnam founded Marietta and, in 1803, Ohio became the seventeenth state.

The ragtag flotilla of vessels on the Ohio River greatly increased following the opening of the Northwest Territories to settlement. Behind this seemingly unplanned and disorganized movement of people on land and (especially) rivers was a widely shared notion of internal improvements. The idea became a significant movement in the first century of the new republic. Its origins, however, had roots in the eighteenth-century colonial era. Notable amongst the promoters of canals and turnpikes was George Washington, who championed the Potowmack and James River and Kanawha Canals and other trans-Appalachian improvements.[6] This movement sought to provide a means for exploiting the rich natural resources of the territory beyond the Appalachian Mountains by a system of roads, canals, and later railways connecting the east-coast ports and industrial centers with the Great Lakes and the Ohio Valley.

Many credit Albert Gallatin's report to Congress in 1808 as the beginning of the internal improvements movement, as it was an attempt to involve the federal government in the development of transportation in the new country. Despite champions such as Jefferson, Washington, and Gallatin, it was not until the granting of federal lands for railways (and in the case of Ohio, the Ohio and Erie Canal) that Congress became involved in supporting internal improvements. In the canal era it was left to states, municipal authorities, banks, and individuals to provide the wherewithal for building and operating canals.

The lack of federal support for internal improvements, except for the National Road, during the early part of the nineteenth century did not deter the entrepreneurial spirit. Each East Coast city sought a means to connect itself with the heartland of the Midwest. New York led the way with the highly successful Erie Canal, completed in 1825. Not to be outdone, Pennsylvania launched the Pennsylvania Mainline Canal, which featured the spectacular railway inclines at Holidaysburg that carried canal boats over the summit of the Alleghenies. Pittsburgh was at last linked to Philadelphia when the canal was completed in 1835, after nine years of struggle. With no direct link to the western waters, Baltimore—in a risky venture—threw its support behind the untried railroad. At the same time, on July 4, 1828, President John Quincy Adams turned the first shovel, marking the beginning of the Chesapeake and Ohio Canal, which was intended (as the name implies) to connect the Chesapeake Bay with the Ohio River at Pittsburgh. On the very same day, ceremonies took place in Baltimore marking the beginning of construction of the Baltimore and Ohio Railroad, which shared the very same objectives. The rivalries between these alternative transportation systems in the Potomac Valley became intense, but the Baltimore and Ohio Railroad reached the Ohio River on Christmas Eve 1852, while the Chesapeake and Ohio Canal languished at Cumberland. Rather than being a main artery to the West, the Chesapeake and Ohio Canal became an important regional canal, hauling coal, timber, and other raw materials to the Washington area and sending manufactured goods as far inland as Cumberland.[7]

Not to be outdone by competition from Northern canals, the Commonwealth of Virginia embarked upon the James River and Kanawha Canal. Even a casual view of a map of Virginia indicates the great advantage of a water route to reach the Ohio River well below Pittsburgh. With a more southern climate, such a route would be ice-free for most of the year, especially since the New River, a principal tributary of the Great Kanawha, rises in North Carolina and brings comparatively warm water into the Appalachian Mountains. Work began on the Kanawha in 1820. With the state company reorganized as a private venture, in the next two decades nearly 200 miles of canal were completed to Cowpasture River, not quite halfway on the projected route. The Civil War put an end to the first phase of construction. Following the war, a new vision, which received endorsement from the federal government, was the Central Water Line stretching from the Chesapeake Bay to the foothills of the Rocky Mountains. This grand improvement was never built, but even in the twentieth century there were proponents of such a waterway. It would have combined a canal, canalization of both the Great Kanawha and Ohio Rivers, and, farther west, the Missouri River to the foothills of the Rockies.[8]

Even before construction began on the great public works to link the East Coast with the Great Lakes and the Ohio Valley, there were men who envisioned connecting Lake Erie with the Ohio River. Amongst the earliest recorded were Jefferson and Washington in 1784, discussing connecting Lake Erie with the Ohio River as part of a much larger scheme to link the Ohio-Mississippi system with the Atlantic Ocean and St. Lawrence River (see fig. 1.1 and fig. 1.2).[9]

In 1807, Senator Thomas Worthington of Ohio offered a resolution in Congress directing Albert Gallatin, secretary of the Treasury, to report on federal aid for canals and roads. The Gallatin Report of 1808 has been heralded as a milestone in the internal-improvement movement, presenting as it does a grand scheme for a national transportation system.[10] It was not implemented on the federal level, and the nation missed an opportunity to develop a rational national transportation system. This left road, canal, and later railway construction to state and local governments as well as private enterprise. Rich in lands but poor in cash, the federal government—later, during the railway mania at midcentury—provided generous land grants

to railway companies, most notably the Central Pacific and Union Pacific Companies, to build the first transcontinental railway.[11] While failing to secure federal aid for roads and canals, the Gallatin Report rose above mere commercial considerations and promoted internal improvements as a means of establishing a more perfect union of the various states.

Without federal leadership and aid, the transportation system that developed was highly competitive and uncooperative, resulting in an irrational system with many canals and local railway lines closing after limited operation. The Erie Canal, organized in 1810 under the leadership of DeWitt Clinton, had a profound influence on other canals developed in America. After failing to receive federal aid, the Erie Canal Company sought to strengthen its renewed appeal by asking the state of Ohio to approve a resolution urging federal support for a project of national importance.[12] It would clearly benefit Ohio, as goods and passengers from the East could be brought up the Hudson River, cross the Erie Canal to Lockport on Lake Erie, and then be shipped to the Ohio ports bordering the lake as well as possibly linking to the Ohio

Figure 1.1. The notable *US Western Engineer*, built by Stephen H. Long of the Corps of Engineers. She sailed in 1818–19 on the lower Mississippi.

River by future canals in the East. There were, however, legal implications: Could the federal government be involved in these great public works, or were they strictly the responsibility of individual states? President Madison thought it was unconstitutional and would not support such initiatives. The War of 1812 effectively ended any chance of federal aid.

Following the War of 1812, DeWitt Clinton again sought aid from Ohio. In response to the initiatives of Governor Worthington, the Ohio General Assembly passed a resolution in 1816 supporting the Erie Canal—but struck the pledge of financial aid from the bill. In the same year Ethan Allen Brown, later governor of Ohio, wrote to DeWitt Clinton on the possibility of a canal linking Lake Erie with the Ohio River. In his inaugural address of December 14, 1818, Brown stated:

> To increase industry and develop our resources, internal communications must be improved to provide for the surplus produce of our state a cheaper way to market.[13]

Figure 1.2. The steamboat *Enterprise* is one of the earliest steamboats. (C. W. Sutphin collection)

The governor strongly urged the case for a canal and that it should be built by the state of Ohio and not by private enterprise. Various bills were presented to the general assembly and debated over the next several years, but no action was taken. It was not until January 1822—after reports, bills, and lobbying of the legislature, that the assembly finally passed the bill authorizing the government to appoint a commission and employ an engineer. The total amount was a paltry $6,000, but at least the project was underway. David S. Bates was appointed the chief engineer together with three commissioners. The report took three years to complete, with five routes examined. These were: 1) Mahoning and Grand River route, 2) Cuyahoga and Muskingum (Tuscarawas Branch), 3) Black and Muskingum (Killbuck Branch), 4) Scioto and Sandusky, 5) Maumee and Great Miami. The commissioners submitted their report in January 1825, urging that a canal be built. Although the project would run on borrowed money, they were confident that the investment would be repaid from revenues by no later than 1837.[14]

By February 4, the general assembly of Ohio had passed an act that outlined the construction of not one but two canals: the Ohio and Erie Canal from Cleveland on Lake Erie to Portsmouth on the Ohio River, and the Miami and Erie running from Toledo on Lake Erie to Cincinnati on the Ohio River. Funds would be raised by borrowing, at six percent, $400,000 in 1825 and an amount not to exceed $600,000 per annum in subsequent years. Appropriation by the general assembly was meant to defray interest charges after any profits were deducted. The vision was to have a canal running diagonally across the state from Cleveland to Cincinnati. From an engineering point of view, crossing the watershed at Portage near Akron posed little difficulty in providing water at the summit. This was not the case in crossing from the Scioto River watershed to that of the Miami River. In fact, the single-canal concept had to be abandoned since the divide was notably higher than the sources of either river, necessitating consideration of alternate routes. Beginning in the west, the Miami and Erie Rivers link the Great Miami River below the divide of the Ohio-Mississippi drainage basin and the Great Lakes watershed with the Maumee River above the divide. Such a route would connect Cincinnati and Toledo. The possible central route would follow the Scioto from Portsmouth to Columbus in the middle of the state, intending to cross the divide, with the northern portion utilizing the Sandusky River to carry the canal to Sandusky-on-the-Lake. The eastern of these three routes crosses a large part of the Muskingum watershed. It would begin in Cleveland and follow the Cuyahoga River from the lake to the summit near Akron, then cross the divide and follow the Tuscarawas branch of the Muskingum as far as Coshocton. In order to stimulate development in the south-central portion of the state, the canal would swing southwest using the Licking River, a tributary of the Muskingum, below Newark and thence follow Walnut Creek into the watershed of the Scioto. The Scioto Valley provided the route to Portsmouth. In an effort to stimulate economic development in the central part of the state, which would garner popular support for this great public work, the canal used the valley of the Scioto rather than following the Muskingum River to Marietta on the Ohio River, which would have been shorter and a better economic solution.

Construction was officially begun on July 4, 1825, three years before the July 4 beginning of the C&O Canal and the B&O Railroad. To show the link with the Erie Canal and Ohio's earlier support of the great public works under DeWitt Clinton's administration, ground was broken by Clinton himself at Newark on the Ohio and Erie Canal; then the ceremony was repeated farther west for the Miami Canal at Middletown, not far from the Indiana border. The earlier Act of 1822 for a survey of the canal on a statewide basis was headed by Micajah Williams from Cincinnati and Alfred Kelley of Cleveland. They enjoyed the sobriquet of "The Fathers of the Ohio canal system" in recognition of the pivotal role they played in building the two principal trans-Ohio canals.

Drawing on the experience gained on the highly successful Erie Canal, three experienced engineers from "Clinton's ditch" (James Geddes, David S. Bates, and William Price) undertook the survey that served as the basis of the 1825 Act.[15] This act established the line of the Ohio (i.e., Ohio and Erie Canal) along the so-called Muskingum and Scioto route.[16]

Several weeks after the passage of the Canal Act, the canal commissioners' annual report, dated February 1, 1825, published "a form of an agreement" intended to provide the specifications for state-sponsored canal construction.[17] Specifications for various canal structures, as stipulated in the agreement, were:

1. The canal alignment would be cleared of trash and trees grubbed for a width of 60 feet to provide sufficient width for the canal prism, berm, and towpath.

2. The canal prism would be 40 feet wide at the water surface, 26 feet wide at the bottom, and 4 feet deep. All of these dimensions were to be the minimum. In many cases the canal had greater prism dimensions.

3. The towpaths would be at least 10 feet wide and smooth and even. They would be pitched so that the path adjacent to the water would be 6 inches higher than the offside to provide drainage.

4. The berm or the side opposite the towpath would not be less than 5 feet in width.

5. The lock chamber would be 90 feet in length and 15 feet wide in the clear. (Note: this a standard size used on the Erie Canal, the Chesapeake and Ohio Canal, and others.)

6. The walls were to be solid masonry laid in water cement (i.e., hydraulic cement, which sets under water and is waterproof) and graded and grouted with water cement mortar at least every 2 feet of height. The walls were to be 5 feet thick at the bottom of the lock and 4 feet at the waterline of the upper level. Buttresses connected with the land wall would rise from the bottom of the wall to the top waterline. Placed 12 feet apart, these buttresses were to be 4 feet in length and extending 4 feet back from the face of the wall.

7. Lock gates together with paddle gates (wickets) and miter sills were to be constructed to plans furnished by David S. Bates, principal engineer.

8. Culverts were to be built in a substantial and workmanlike manner at places designated on the plans and of such dimensions as specified. They were to be built of good, substantial stone laid up in hydraulic mortar.[18]

The above items, based upon the agreement, give insights into the construction details of the locks and dams that made up the Ohio and Erie Canal.

An amendment to the original Canal Act, dealing with the means and the consequences of connecting the Ohio and Erie Canal with the upper reaches of the Muskingum River, was enacted by the General Assembly of Ohio on February 11, 1828.[19] (The story of the Muskingum improvement will be dealt with in the following chapter.) Although improving the Muskingum for navigation was undertaken later, the Ohio and Erie Canal would benefit from river commerce developing on the Muskingum, which was also an effective means of disposing of surplus water in the reaches of the canal below the divide and above Coshocton.

The Board of Commissioners' report on the work accomplished during the calendar year 1828 was published on January 6, 1829.[20] A year earlier it was anticipated the canal would be open in May for navigation from Akron to Massillon, followed a month later by the section from Massillon to Dover. Construction was considerably delayed because of unprecedented rainfall that caused widespread flooding in the Muskingum watershed. The summit lake at the Massillon section was delayed until the end of August, whereas the lower section, Massillon to Dover, was completed in June 1829. Coupled with rain, a shortage of labor developed that the commissioners attributed to intense construction work on internal improvements in Pennsylvania.[21] Although of a temporary nature, the delay in completing the canal and putting it in service from Portage Summit at Akron to Newark, where the canal left the Muskingum watershed and crossed a divide to the Scioto, postponed any toll receipts that were expected to offset, in part, the interest on the borrowed construction money.

The wet summer yielded to a fine, dry autumn and construction work was pursued with vigor. The turn in the weather allowed the masonry aqueduct over the Tuscarawas River near Bolivar to be completed. The commissioners were obviously well pleased, since they believed the aqueduct to be "equal in strength and beauty to any work of this kind in the United States."[22] Two other aqueducts, near Newark over the Raccoon and North forks of the Licking River, were also completed except for the planking of the "trunk" carrying the water. The dam thrown across Licking Narrows joined the list of completed projects.

In the Muskingum watershed, the work was advancing steadily with lock walls completed at a total of twenty-one sites between the Licking Summit and Caldersburgh as well as between Massillon and Dover; two additional locks were nearing completion. As in the previous year, when the board expected the opening of two sections in the Muskingum watershed, the commissioners expected the entire division to have water let in by July 1, weather permitting.[23]

The great floods of January 1829 dealt a severe blow to the region, especially that portion north of the Portage Summit. These floods caused a rise of more than two feet above any previous records on the Cuyahoga, and yet the canal itself sustained only limited damage, requiring less than $6,000 to repair. Nevertheless, countermeasures were undertaken to mitigate future flood damage. This work included raising levees and creating extensive waste ways, formed by lowering the towpath in selected places while, at the same time, providing protection of these channels against erosion.[24]

The experience of a year of boating on selected sections of the canal called for several improvements to the locks. It was found that careless or unskilled navigators on the canal had caused damage to the locks and, needless to say, to the banks themselves in attempting to enter a given lock. This was the most difficult part of navigating a boat on the canal. The solution, which is still a current practice, was to build training walls extending beyond the lock chambers in line with the landside lock walls, making maneuvering into the lock much easier. In the case of the Ohio and Erie Canal, these training walls were timber frames built both upstream and downstream of the locks and supported on strong timber piles that would sustain collisions with the canal vessels. In addition, regulating weirs and conduits allowed surplus water to pass around a given lock. These improvements enabled the commissioners to dispense with regular locktenders and allowed crews on the boats to operate the lock gates. This was not the case for similar locks and regulating weirs on the C&O Canal, where each lock or series of locks had a locktender.[25]

Considering the lack of accurate rainfall data and associated runoff information, a critical part of canal design was to provide adequate water where it was needed at the summit as the canal passed from one watershed to another. At the Portage Summit, there was a sufficient supply and, in fact, more than adequate to supply the needs of the Ohio and Erie Canal even during seasonal low-water levels in August and September.

The canal board made a decision for the course of the canal along the Scioto River from Piketon to Portsmouth on the Ohio River at a meeting convened at Chillicothe on July 10, 1829. The route selected proceeded south along the west side of the Scioto River. One of the compelling arguments noted was that the land on the east side would make construction necessary through rough ground and thus incur greater expenses. The opinion, strongly stated by the board, read "they reported that leveling only at an incurred cost and risk, altogether incompatible with the public interest, and unwarranted by any advantage resulting for the measure." Another point offered by the board added to the argument: "The hazard of crossing a river as large as the Scioto, after its floods have been augmented by the waters of all its principle tributaries, forms a strong objection to the adoption of the line on the east side of the river."[26]

In an expansive mood, the board declared in its 8th Annual Report, dated January 9, 1830, that the northern division of the Ohio and Erie Canal from Cleveland on Lake Erie to the north end of the deep cut that took the canal into the Scioto Valley was nearly complete except for some minor work between Dover and Calderburgh. This work required but a few days to complete. Thus water could be let in nearly 190 miles of canal, including all of the path in the Muskingum watershed. After testing of various sections of the canal with a moderate amount of water, only minimum work was required on the Tuscarawas and Walhonding feeders to make them serviceable.

The entire line from Licking Summit to Portsmouth on the Ohio River, a distance of 119 miles, was under contract, the commissioners were pleased to announce. The contracts called for overall work to be completed by June 1, 1831. The canal board's report said that "the locks, aqueducts, and other important structures on the canal, have so far, fully answered the purposes for which they were designed, and that no serious injury has been sustained by any of them since their completion."[27] The sixty miles of canal work from the Licking Summit to Deer Creek, eight miles above Chillicothe, was complete. The remaining fifty-nine miles to the Ohio River were under contract, but little work at that stage had been completed.

The stretch of canal from Massillon to the lake opened for boating during 1829. It had performed well, requiring less than $500 to repair breaches and some landslips. The Board of Commissioners and its engineers had learned much about canal engineering techniques for future work.

As the canal neared completion, the canal board dispensed with the services of the principal engineer, David Bates, on February 1, 1829, believing that the remaining work could be handled by the resident engineers in the various divisions. They were not unhappy with Bates's work; the action was taken to save costs.[28]

In all large public works, there are claims by the contractors for extra compensation or for an extension of the contract. The Ohio canal systems were no exception. The canal board had to deal with numerous claims for increased compensation by contractors working all along the canal. These claims focused principally on the following issues:

1. Work proved more difficult and expensive than anticipated when the contract was signed.

2. Delays blamed on the weather, particularly rainy seasons and floods, which the contractor believed required additional time without penalty.

3. Unanticipated increases in labor and material costs.

4. Changes in the plan, including the location of the canal.

In the case of damage sustained by floods, the canal board would negotiate with the contractors. A change order, or indeed a new supplementary contract, would be let for changes in location or in scope of work. All other claims would be rejected. The canal board would also compensate land owners for damages sustained to their property as a result of canal construction work.[29]

An example of a contractor's claim was that of Rathbone and Larrimure, who submitted a memorial to the Ohio General Assembly seeking additional compensation for work north of the Licking Summit in the Muskingum watershed. They claimed the work represented a different character and description than was contemplated by either the Board of Commissioners or the engineer. Upon investigation, Benjamin Tappan, president of the canal board, replied that "there is no evidence to warrant the conclusion, that the work on the said canal, on the whole, proves to be on the character materially worse than was supposed by the Commissioners and the Engineer."[30]

From Lake Erie to the Deep Cut at Licking Summit, a distance of 190 miles, the total construction cost amounted to $2,005,451.21 (an average of $10,555 per mile).[31] This constituted all the construction in the Muskingum watershed, the canal from Portage Summit to Lake Erie (including the price tag for the great reservoir at Licking Summit—today's Buckeye Lake), as well as non-navigable feeders and miscellaneous other contributing structures.

According to the eleventh report of the canal board, the work in the Scioto Valley was so far advanced that with a favorable construction season the work would be completed in 1831. The Chillicothe and Portsmouth division included the masonry foundations of the aqueducts and large arches that had been completed, together with much of the stone and masonry above the foundations. The foundations of the large dam across the Scioto River, seven miles below Chillicothe, were in place, and the masonry work for constructing the crib dam was at hand. Several locks were finished and the foundations laid for nearly all of the remaining locks. The embankments, which were a particularly visible feature of the route of the canal, were also well advanced in the Scioto Valley. The commissioners anticipated that the entire canal would be opened to traffic in the spring of 1832. Newark saw the first canal boat to traverse the canal from Cleveland on July 10, 1830:

> The first boat which passed from Cleveland, to Newark, arrived at the latter place on the 10th day of July last. The navigation between Dover and Newark, could not however be considered permanently established, and regular until about the first of September. Since that time the commerce on the northern division of the canal, has been active and subject to little interruption, from breaches or other causes; and from the early part of October, until the closing of the navigation for the season, no breach has occurred.[32]

The canal board proudly announced in the report to the general assembly January 22, 1833, that their charge embodied in the Act of 1825 had successfully been carried out, with the exception of work at Portsmouth and locks connecting the Miami Canal with the Ohio River at Cincinnati. These two projects were finished and open for traffic in 1834. In the case of the Ohio and Erie, the following details were presented in the report:[33]

Names of Canals and Branches	Length— Miles and Chains	Lift Locks No.	Rise and Fall. Feet and Hundredths	Guard Locks No.	Aqueducts	Culverts	Stone	Wood	Draws for Crossing
Ohio Canal—Main Trunk	308.14	146	1,207.35	5	14	153	50	8	6
Tuscarawas Feeder	3.20	. . .	. . .	1	. . .	. . .	. . .	. . .	1
Walhonding Feeder	1.30	. . .	. . .	1	. . .	. . .	. . .	. . .	1
Granville Feeder	6.14	1	10	1	1	1	3	. . .	1
Muskingum Side Cut	2.58	3	28.79	. . .	1	. . .	. . .	. . .	. . .
Columbus Feeder	11.60	2	13.90	1	. . .	1	2	1	1
Total–Ohio Canal	33.36	152	1,260.04	9	16	155	55	9	10

The 1822 survey of Ohio canals provided the general assembly with an estimate of $3,081,880.83 for the entire project, whereas the total expenses as of December 1, 1832, after the canal construction was completed, was $4,244,539.64—an apparently very large percentage overrun. The scope of the work, however, changed during construction, greatly increasing the final total expenditure. In evaluating the cost, the commissioners stated:

1. There was additional work, including the Muskingum Side Cut and the Walhonding feeder.

2. Additional aqueducts were constructed over the Walhonding at Roscoe and the Scioto at Circleville on the lower reaches of the river.y

3. Levees were built to give flood protection in selected areas of the canal.

4. Alterations to serve a local commercial interest, especially moving the canal from the east side to the west side of the Scioto River in Highland, Fayette, and part of Ross Counties.

5. Remedial work on portions of the canal after the contractors had completed their work. This work was necessary before the canal could be placed in service.

6. Ordinary maintenance on the canal from 1827 to 1832 when only portions of the canal were in operation and thus providing limited revenues.[34]

In the present age of massive cost overruns on public-works projects, the $4.25 million for the Ohio and Erie Canal construction, taking into account the expanded scope of work compared to the project envisaged in the survey undertaken from 1822 to 1825, seems both appropriate and reasonable.

Proponents of the national internal-improvement movement sought to connect the Ohio and Erie Canal with both the Pennsylvania Main Line Canal and the Chesapeake and Ohio Canal, expecting each of these to terminate in Pittsburgh. Such a link would allow waterborne commerce from Ohio to reach the East Coast. Incorporated by a private company in 1828, the Sandy and Beaver Canal connected Bolivar on the Ohio and Erie Canal in Stark County, Ohio, with the Little Beaver River, which joined the Ohio at East Liverpool only forty miles below Pittsburgh. The construction work on the twenty-three-mile waterway was not begun until 1834. Caught in the financial panic of 1837, the work was stopped and not resumed until 1845. With a challenging tunnel to construct across the divide between the watersheds and a lockage of 669 feet, the work was not completed until 1850. By this time, railway competition was intense, resulting in very little traffic on the Sandy and Beaver. It was in service a mere two years. The only portion in operation beyond that time was the twelve-mile Nimishillen link serving Canton. The state took over control in 1856.[35]

A second lateral canal, the Pennsylvania and Ohio, proved to be more successful. Incorporated in 1827, construction began in 1836 by a private company, with state aid, to connect Akron at the Portage Summit of the Ohio and Erie Canal with the Pennsylvania Main Line Canal and thence into Pittsburgh on the Ohio River. From Akron, the canal passed through Kent where there is a notable surviving masonry-arch dam and associated lock, thence to Ravenna Summit, where it passed into the Mahoning watershed and followed that river by Warren and Youngstown for a total length of approximately ninety-three miles. Finished in 1840 to Pittsburgh, it was completed late in the canal era and at the height of the railway "mania." Both of these lateral canals had little prospect of success.[36]

Two other branch canals were built as part of the Ohio and Erie system. The Walhonding feeder ran twenty-five miles up the Walhonding Valley toward the center of the state in anticipation of transporting farm produce from this rich agricultural region. To make the feeder possible for navigation, eleven locks and dams with a lockage of eighty-five feet, together with two guard locks, were built on the original feeder canal. Authorized in 1836, the navigation opened for traffic in 1841.[37]

Much farther south in the Scioto watershed, the Hocking Canal formed a fifty-six-mile branch of the Ohio and Erie Canal. Built in three sections, the entire waterway opened in 1843. The first section was constructed by a private company incorporated in 1826 under the name of the Lancaster Lateral Canal, while the remainder was built by the state of Ohio. The sixteen-mile section from Lancaster toward Athens was put under contract in 1836, followed the next year by a stretch to Nelsonville; the remaining section was begun in late 1838. At the same time, the state also purchased the Lancaster Canal, completed to Athens in 1843. There was talk of extending the canal to Hockingport on the Ohio River, but this idea languished, and the canal was destined to be a local feeder to the Ohio and Erie Canal, serving agricultural interests in the Hocking Valley as well as coal and salt industries stretched along its length. Like the lateral canals to the north, neither of these branches was a financial success.[38]

Ironically, the Ohio and Erie Canal network produced a net income of seven million dollars in the antebellum period and saw a peak year in 1851, while it became increasingly clear that the railways were superseding towpath canals in many parts of eastern America. The 1850s saw an unprecedented expansion of the railways to the detriment of canal traffic. Apart from speed of delivery, the railways had other decided advantages. Towpath canals, even with mechanical means of towing, were restricted to about four miles per hour because wake at higher speeds caused damage to the canals. Paddle-wheel and screw propulsion were equally damaging. Once a right-of-way was secured, railway track could be laid quickly and cheaply in contrast to canal building. In operations, railways did not have to curtail operations in the winter because of ice and rarely from snowstorms. It could, however, still be argued that for shipping bulk cargo such as coal, canals provided the cheapest cost per ton/mile.

The canals in the United Kingdom, which began with the Bridgewater Canal in Lancashire in the middle of the eighteenth century, dominated the transportation system of that country, forming the arteries of the Industrial Revolution that remained unchallenged until the railway age in the late 1820s. The British system of canals flourished for nearly three-quarters of a century. The canal age in America, in contrast, was compressed into only several decades before railways became an economic threat by the 1840s. In fact, the Ohio Canal's enabling legislation of 1825 was only three years in advance of the ceremonial beginning of the Baltimore and Ohio Railroad in 1828. The railways had built increasing lengths of track in Ohio by midcentury, and political leaders, businessmen, and the public in general turned increasingly from being promoters of canals to enthusiastic entrepreneurs for the new technology.

It was more than just talk. The options were to limit investment in the canals and let them deteriorate, lease the system to private enterprise, or get out of the canal business completely and sell the entire Ohio Canal system. After much discussion, an act of May 8, 1861, provided for the state-owned public works to be leased for ten years, the lease subsequently to be extended for another decade.[39]

Although the act required the lessees to maintain the system in good repair, it failed to provide a means of establishing the condition of the canals and associated structures at the time the lease was signed. It proved impossible then to determine what responsibility for repairs and maintenance rested with the new renters. This extensive system rented for a mere $20,075 per annum. A clear intent for leasing the system was, however, stated by Governor Young when he said, "so as to keep said Public Works in good and proper condition of repair and deliver up said Public Works to the State of Ohio at the termination of said lease, in like good condition."[40] After the city of Hamilton filled in part of the canal basin, the lessees refused to pay their half-yearly rent. As a result, the Board of Public Works terminated the lease. It was also during this period that a number of unprofitable canals were abandoned on a national basis.[41]

In lieu of abandonment, the federal government entered the scene in a meaningful way. There was renewed interest in the government taking over existing canal projects. These included the canalization of the Big Sandy River that forms the border between West Virginia and Kentucky, taking over the Little Kanawha Navigation from a private company, and building an additional lock, designated the "government lock" on that river.[42] A similar action resulted in the Muskingum Improvement becoming a federal waterway in 1887.[43] By far the most ambitious and indeed successful project was the complete canalization of the Great Kanawha River from the falls to the river's confluence with the Ohio River at Point Pleasant. This work was begun in 1873 and completed in 1898 with eleven locks and movable dams.[44]

These canals and river navigations were just part of a vision to improve navigation on the Ohio River with a series of locks and French-style movable dams stretching from Pittsburgh nearly 1,000 miles to the confluence of the Ohio and the Mississippi Rivers at Cairo, Illinois. The first lock and dam (designated No. 1) at Davis Island was begun in 1878; it was the first of more than fifty such structures along the entire length of the Ohio River. The navigation was not completed until 1929.[45]

Unlike towpath canals with restricted dimensions and limited speed, river navigation on the Ohio and Great Kanawha proved to be a vital network in the nation's transportation system. Rates for hauling bulk cargos were unchallenged by any other system.

To overcome the limitations of towpath canals, the U.S. Congress empaneled a commission in 1894 to undertake a feasibility study for building a seven-foot-deep barge canal across Ohio to link Lake Erie with the Ohio River. They reported in 1896, having determined that there were three natural routes. The eastern route comprised the Muskingum Navigation (already complete) and improvement of the northern division of the Ohio and Erie Canal from Coshocton to Cleveland. The central route, with the least lockage and shortest distance, would be the most economical to operate, but the route lacked port facilities at either Sandusky on Lake Erie or Portsmouth on the Ohio River. The western route from Cincinnati to Toledo via Dayton, utilizing the Miami and Maumee Rivers, would be the most expensive to build, but it would serve an industrial corridor and enjoy developed terminal facilities at both the lake and at the Ohio River.

The eastern route, including the northern division of the Ohio and Erie Canal together with the improved Muskingum Navigation, was the least favorable regarding both the estimate for future traffic and water supply across Portage Summit near Akron. In its defense, the commissioners projected that such a route would have large local traffic in coal, clay, limestone, and agricultural products. Because of improvements already in place on the Muskingum, the eastern route was estimated to be significantly cheaper than the other two routes. The commissioners reported that it was feasible not only for a seven-foot-deep channel, but for one of ten feet in depth. This would allow the largest boats plying the Ohio River system and the smallest ships that could safely navigate the Great Lakes to use this eastern route.[46] The following estimates were presented:[47]

As to the cost of a ten-foot barge canal, the commission reported:

Eastern route	$15,042,586
Central route	20,784,451
Western route	26,865,126

By definition, a barge canal would employ steam-hauled tows or self-propelled boats. While supporting the feasibility from an engineering point of view and the undoubted benefits to the state of Ohio, the report did not believe that the United States government should construct such a route. In fact the scheme was never implemented, resulting in the U.S. Army Corps of Engineers being responsible in Ohio for the Muskingum navigation only, in addition to the extensive system of locks and dams on the Ohio River itself. It appears that the reasoning in the report was that such an improvement, although of benefit to the state of Ohio, was not in general a benefit to the country as a whole.

The role of the canals in the industrialization of Ohio, from a purely economic point of view, was summarized in the following table:

RECORD OF THE CANALS

Cost of construction	$15,967,652.69
Cost of maintenance and operation to Nov. 15, 1903	12,063,849.47
Total cost of canal system	28,031,502.16
Gross receipts 1827–1903	16,953,102.98
Deficit	11,078,399.18

Present value entire property estimated at	15,000,000.00
Balance in favor canals	3,921,600.82
Gross receipts as shown above	16,953,102.98
Cost of maintenance and operation	12,063,849.47
Net revenue to the state not including interest on loans, cost of construction, or recent value of canal property	4,889,253.51[48]

This table traces the economic history of the canal from the beginning of construction to 1905. Only the Muskingum Navigation continued in operation until the 1960s. The principal canals of Ohio played a very limited role in the economic life of the state during the twentieth century. The improvement of the Muskingum River needs to be considered as an attempt to control the water resources of the entire watershed for navigation purposes.

NOTES

1. U.S. Army Corps of Engineers, Ohio River Division. *Ohio River Navigation: Past-Present-Future* (Cincinnati, OH: The Division, U.S. Army Corps of Engineers, 1979), 3–4.

2. Michael C. Robinson, *History of Navigation in the Ohio River Basin* ([Fort Belvoir, VA]: National Waterways Study (NWS-83-5), U.S. Army Resources Support Center, Institute for Water Resources, 1983), 2–3; Leland R. Johnson, *Men, Mountains and Rivers: An Illustrated History of the Huntington District, U.S. Army Corps of Engineers* (Washington: GPO, 1977), 2–3.

3. Johnson, *Men, Mountains and Rivers*, 3.

4. Martin R. Andrews, *History of Marietta and Washington County, Ohio* (Chicago: Biographical Publishing Co., 1902), 215–216.

5. U.S. Army Corps of Engineers, *Ohio River Navigation*, 3–6.

6. Ronald E. Shaw, *Canals for a Nation: The Canal Era in the United States 1790–1860* (Lexington, KY: University of Kentucky Press, 1990), 7–9.

7. Ronald E. Shaw, *Chesapeake and Ohio Canal* (Washington: National Park Service, 1991), 4–40.

8. Wayland F. Dunaway, *History of the James River and Kanawha Company* (New York: Columbia University Studies in History, Economics and Public Law, vol. 104, Columbia University Press, 1922); Emory L. Kemp, *The Great Kanawha Navigation* (Pittsburgh: University of Pittsburgh Press, 2000), 5–14.

9. Shaw, *Canals for a Nation*, endnotes 9 and 15.

10. Albert Gallatin, *Report of the Secretary of the Treasury on the Subject of Public Roads and Canals, 1808* (New York: Reprints of Economic Classics, Augustus M. Kelley, 1968).

11. Stephen E. Ambrose, *Nothing Like It in the World: The Men Who Built the Transcontinental Railroad, 1863–1869* (New York: Simon and Schuster, 2000).

12. C. C. Huntington and C. P. McClelland, *Ohio Canals, Their Construction, Cost, Use and Partial Abandonment* (Columbus, OH: Ohio State Archaeological and Historical Society, 1905), 9.

13. Huntington and McClelland, *Ohio Canals*, 10.

14. Shaw, *Canals for a Nation*, 127–130; Huntington and McClelland, *Ohio Canals*, 10–15.

15. Shaw, *Canals for a Nation*, 38.

16. John Kilbourne, *Public Documents Concerning the Ohio Canals* (Columbus, OH: I. N. Whiting Publishers, 1832), 158–165.

17. Kilbourne, *Public Documents*, 212–216; Huntington and McClelland, *Ohio Canals*, 152–162.

18. Kilbourne, *Public Documents*, 212–216; Huntington and McClelland, *Ohio Canals*, 152–162.

19. Kilbourne, *Public Documents*, 317–319.

20. Kilbourne, *Public Documents*, 323–333.

21. Kilbourne, *Public Documents*, 323.

22. Kilbourne, *Public Documents*, 324.

23. Kilbourne, *Public Documents*, 353.

24. Kilbourne, *Public Documents*, 325–326.

25. Kilbourne, *Public Documents*, 356.

26. Kilbourne, *Public Documents*, 331–332.

27. Kilbourne, *Public Documents*, 356.

28. Kilbourne, *Public Documents*, 357.

29. Kilbourne, *Public Documents*, 362–365.

30. Kilbourne, *Public Documents*, 405.

31. Kilbourne, *Public Documents*, 407.

32. Kilbourne, *Public Documents*, 449.

33. Huntington and McClelland, *Ohio Canals*, 31.

34. Huntington and McClelland, *Ohio Canals*, 32–34.

35. Huntington and McClelland, *Ohio Canals*, 38–41, 48–49.

36. Huntington and McClelland, *Ohio Canals*, 40.

37. Huntington and McClelland, *Ohio Canals*, 40.

38. Shaw, *Canals for a Nation*, 131, 134.

39. Huntington and McClelland, *Ohio Canals*, 47–48.

40. Huntington and McClelland, *Ohio Canals*, 47–48.

41. Huntington and McClelland, *Ohio Canals*, 48–51.

42. Larry N. Sypolt and Emory Kemp, "The Little Kanawha Navigation," *Canal History and Technology Proceedings*, vol. X, 1991 (Easton, PA: Canal History and Technology Press, 1991), 41–93.

43. Johnson, *Men, Mountains and Rivers*, 100.

44. Kemp, *The Great Kanawha Navigation*, 42–81.

45. William E. Kreisle, "Development of the Ohio River for Navigation 1825–1935" (M.A. thesis, University of Louisville, 1971), 140–142.

46. Huntington and McClelland, *Ohio Canals*, 139–141.

47. Huntington and McClelland, *Ohio Canals*, 140.

48. Huntington and McClelland, *Ohio Canals*, 110.

CHAPTER TWO

Slackwater Navigation on the Muskingum

T he riverboat *New Orleans* inaugurated steam navigation on the Ohio
River in 1811 with a memorable voyage to New Orleans from Pittsburgh. The boat had been built by Nicholas J. Roosevelt for the Fulton/
Livingston partnership, which, having already pioneered steam navigation on
the Hudson River, now hoped to control steamboat traffic on the Western
waters.[1] A few years later, in 1816, the steamboat *Elisha* arrived in Charleston—
the first steamboat to navigate the unimproved Great Kanawha River from
its confluence at Point Pleasant.[2] With these exploits in mind it is little won-
der that interest mounted along the Muskingum for the possibility of steam
navigation on that river to open up the Muskingum Valley for development,
both agricultural and industrial. Most appropriately, the steamboat named
for the founder of Marietta, Rufus Putnam (see fig. 2.1), cast off her moor-
ings at Marietta and headed for Zanesville, some seventy miles upstream.
With the river running at or near flood level, no obstructions were encountered, and the gallant little steamboat reached Zanesville that evening after
twelve hours steaming against the current. The date was June 9, 1824, two
years after approval for a survey of possible locations for canals across Ohio.[3]
Despite the highly visible voyage of the *Rufus Putnam* and considerable interest by the inhabitants of the valley, the Muskingum River was not included in
the study nor subsequently in the act of the Ohio General Assembly passed
in February 1824.[4]

Of the two principal routes selected, the Miami and Erie Canal and the
Ohio and Erie Canal connecting Lake Erie with the Ohio River, it was the
Ohio and Erie that utilized the Tuscarawas branch of the Muskingum to
carry the canal from the summit level near Akron to the Licking Summit
at the southwestern edge of the Muskingum watershed. Rather than using

the Muskingum River, however—the shortest route to the Ohio River—the canal turned west, utilizing the valley of the Scioto and the Ohio River terminus at Portsmouth, Ohio.

Without substantial hydraulic improvements in the river, steamboat navigation was not feasible except under unusual river levels. Therefore in 1827

Figure 2.1. Rufus Putnam founded Marietta, Ohio, and named it for Marie Antoinette. (C. W. Sutphin collection)

the Ohio General Assembly authorized the Board of Canal Commissioners to survey the Muskingum Valley for a link between the Ohio and Erie Canal and the Ohio River at Marietta. In compliance with the resolution of the general assembly, the Board of Canal Commissioners reported that "The character of the valley and the channel of the Muskingum, render it much cheaper to make a steamboat navigation in its channel, than a canal along its margin."[5]

The estimate presented for this improvement from the lower bridge in Zanesville to the Ohio River was $353,443.67. In order to connect the Muskingum Navigation with the Ohio Canal, the river would need to be improved from Zanesville to Dresden. If a connection could be effected, any surplus water in the canal between the Portage divide of the Licking Summit could be discharged into the Muskingum River.

The total distance was very close to ninety miles, with a descent of 158.5 feet between Dresden and Marietta or 129.67 feet from the low-water mark in the Ohio River at Marietta to the low-water mark at Dresden. In order to accommodate shallow-draft steamboats, eleven locks and associated dams, with chambers 150 feet long and 34 feet wide, would be required. Unlike the other towpath canals being built, the Muskingum would be a river navigation achieved by a series of slackwater pools. This improvement could be effected by improving the Muskingum to its confluence at Coshocton where it would join the Ohio and Erie Canal. A closer connection, however, could be made with a "Side Cut" from Dresden, a distance of some 2.5 miles being served with a series of three locks. The seventh Annual Report of the Board of Canal Commissioners documented that the "Side Cut" was approved in February 1828.[6]

The board also reported that a lock and dam were to be constructed above Zanesville to provide a slackwater pool permitting vessels to reach Dresden and make the junction with the Ohio and Erie Canal. Extensive work was to be undertaken at Zanesville, consisting of a bypass canal 2.5 miles long with three locks and a dam in the river. The commissioners concluded that constructing a slackwater navigation would require eleven locks and dams to overcome a fall of just over 104 feet from Zanesville to Marietta.

In the same year that the Ohio and Erie Canal was completed in 1832, the Ohio General Assembly authorized the dam between Dresden and

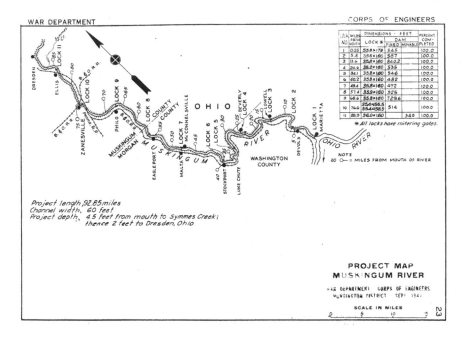

WAR DEPARTMENT   CORPS OF ENGINEERS

L&D NO.	MILES ABOVE MOUTH	LOCK *	DAM FIXED	DAM MOVABLE	PERCENT COMPLETED
1	0.25	358x179	545		100.0
2	5.8	358x160	587		100.0
3	13.6	358x160	840.2		100.0
4	24.6	358x160	535		100.0
5	34.1	358x160	546		100.0
6	40.2	358x160	482		100.0
7	49.4	358x160	472		100.0
8	57.4	359x160	525		100.0
9	68.6	358x160	729.6		100.0
10	76.6	354x56.5 / 354x58.5	514		100.0
11	85.9	360x160		340	100.0

* All locks have mitering gates.

Project length, 92.85 miles
Channel width, 60 feet
Project depth, 4.5 feet from mouth to Symmes Creek;
thence 2 feet to Dresden, Ohio

NOTE
20 O— = MILES FROM MOUTH OF RIVER

PROJECT MAP
MUSKINGUM RIVER

WAR DEPARTMENT CORPS OF ENGINEERS
HUNTINGTON DISTRICT SEPT 1947

SCALE IN MILES

Zanesville at Symmes Creek, 10.7 miles above the Zanesville Dam. On most American canals, the lowest lock and dam was designated No. 1. This was not originally the case for the Muskingum waterway, where the first lock constructed was at the highest level and designated No. 1. Years later, this numbering system was reversed with No. 1 being designated for the lock and dam at Marietta. The numbering system probably changed after the federal government took over the navigation so that some locks are referred to by two numbers.

It appears that the Zanesville hydraulic works recommended by the commissioners were not included in the act of the General Assembly because of proprietary rights held by the Zanesville Canal and Manufacturing Company, which would undertake such works for waterpower potential at the dams. Since the company failed to complete the dams in a timely fashion as required by the assembly, its charter was revoked and the works completed by the state under an act passed on February 19, 1835 (see fig. 2.2).[7]

Figure 2.2. US Army map of the Muskingum River. (Project Maps, US Army Corps of Engineers, 1947)

The scope of work of the Board of Canal Commissioners was expanded to include state-sponsored public works, hence the name was changed to the Board of Public Works (B.P.W.). The act of 1835 authorized the B.P.W. to undertake the construction of river improvements on the Muskingum from Zanesville to its confluence with the Ohio at Marietta, according to the requirements set out in the earlier act of 1827, which recommended eleven locks and dams. The situation was complicated by the general assembly granting private parties rights to construct two dams across the river below Zanesville to provide waterpower. These dams were located at present Philo and Lock and Dam No. 9. A further grant was made to Robert McConner to construct a dam at the present Lock and Dam No. 7.[8]

Before the advent of the pound lock with two sets of gates at either end of a lock chamber, access for vessels wishing to pass through mill dams was provided by a single gate, which could be raised, swung open, or dropped to permit passage of a vessel. When opened, a great surge of water was released, hence the term "flash lock."[9] This medieval system reappeared at the private waterpower dams on the Muskingum. It was clear that these private dams were not suitable for the proposed navigation system. To extricate itself from this situation, the state provided compensation to the owners of the dams. The essence of the arrangement involved the owners forfeiting their right to operate a dam in return for a guaranteed perpetual right to use waterpower at the new lock and dam. Contracted in 1836 for a total of $1,627,018.20, the waterway was completed in 1841 and opened for commerce on October 1.

By extending the navigation along the river above Dresden, the Muskingum waterway could join the Ohio and Erie Canal, providing a connecting link between canal towpath barge traffic and steam navigation on the Muskingum. It was never intended for steamboats to use the Ohio and Erie Canal but rather that the canal traffic could be extended to the Muskingum and provide service to Zanesville and even Marietta. In the other direction, canal traffic could move Muskingum Valley goods to Lake Erie via the Ohio and Erie Canal.

Connecting these two systems seemed highly desirable at the time. The initial idea was to join the Ohio and Erie at Coshocton. With this in mind,

the general assembly issued an act in 1829, entitled "to amend the Act for the protection of the Ohio Canals," which said, in part: "That the Canal Commissions be, and they are hereby authorized and empowered to cut a navigable side cut or branch canal from the main canal, to enter the Muskingum River at or near the town of Dresden."[10]

A cursory review indicated that the link should be made at Coshocton. The report concluded, however, that "a side cut or branch canal of about 2½ miles in length from the main canal to the Muskingum at Dresden, with three boat locks, overcoming a descent of about 28 feet from the canal into the river, will also be necessary to perfect the plan."[11]

Thus, the side cut near the town of Dresden was located and put under contract to be finished one year hence, in 1830. In this short distance, the cut crossed Tomika Creek on an aqueduct and descended to the river at Dresden. According to the January 11, 1831, report of the canal commissioners, the side cut was completed during the previous year.

Figure 2.3. A typical crib dam shown at Lock and Dam No. 10, Muskingum River. The GH Crib Dam features wood planking on top and upstream, and it can be seen that this dam needs to be repaired. (Project Maps, US Army Corps of Engineers, 1947)

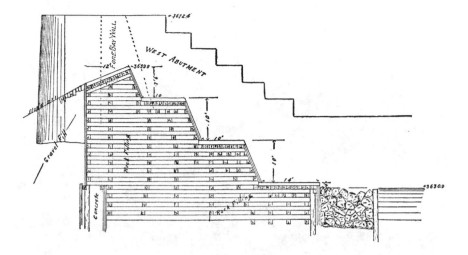

Slackwater navigations suitable for steamboats posed special problems in the design of the locks and the dams. The solutions evolved from the current practice of towpath canals, in which the standard American canal featured lock chambers ninety feet long by fifteen feet wide with a lift of about eight feet.

Crib Dams

The crudest hydraulic structure was the widely used crib dam, which consisted of rock-filled timber cribs, built like an open box reminiscent of loghouse construction. After being filled with stones, the dam was faced with timber planks. In service, these crib dams resisted water pressure as gravity dams. Quick and easy to construct, a crib dam lacked the durability of more permanent masonry dams. Built by the score, timber cribs filled with stones became the American standard for stationary dams for both navigation and waterpower. There were two standard types, both used on the Muskingum, namely, the slope dam and the step dam (see fig. 2.3). In the former, the cribs are decked over with solid timber sheathing installed on a slope (see figs. 2.4 and 2.5), whereas the step dam created a series of cribs diminishing in height as they marched downstream. All of the original crib dams on the

Figure 2.4. A stepped wood crib dam as rebuilt. (*The Design and Construction of Dams* [1911], by Edward Wegman)

Muskingum Navigation were of the slope-dam type, but after a flood in 1837 the dams were subsequently repaired as step dams. The slope dam passed ice and debris more readily than the step dam, whereas under normal river levels, the step dam dissipated the discharge and caused less turbulence at the foot of the dam and farther downstream. In designing crib dams, engineers had to prevent overturning of the dam, the possibility of sliding on the bottom of the river, and the excessive penetration of water between the bottom of the base of the cribs and foundation. As a result, the bottom timbers were usually secured to a rock foundation with bolts or, in the case of foundations of softer material, the bottom members were secured to timber piles driven into the riverbed.

Although cribs dams were occasionally built inside temporary cofferdams, the usual method was to lay the timbers during periods of low water. The

Figure 2.5. A badly damaged crib dam on the Muskingum Navigation. (C. W. Sutphin collection)

preferred wood was white oak or yellow pine, but any of the more solid, heavier wood species were used. Some of the early dams used logs or hewed timbers; later, sawn timbers eight or ten inches square were standard. The sheathing was generally of sawn timber but occasionally hewn planks were used. At low water, at Lock and Dam No. 4 on the abandoned Little Kanawha Navigation, it is possible to see the bottom timbers of the crib dam and the bolts securing them to the rock foundation. Despite design knowledge that would provide engineers with methods for proportioning crib dams, many were built on an empirical basis. Even in the best-built crib dams, leakage was always a problem, as were timbers that were not continuously submerged rotting and requiring replacement.[12]

Later, concrete caps were added by the Army Corps of Engineers at Locks and Dams Nos. 4, 5, 7, and 10. The first ten dams ranged in length from 472 feet at Lock and Dam No. 7 to 840.2 feet at Lock and Dam No. 3, but the length of the concrete Lock and Dam No. 11 measured only 340 feet. In order to preclude flooding upstream, a movable dam of the French Boulé type was installed. With the wicket panels removed, the support frames (fermettes) could be folded flat on the top of the concrete crest of the dam, allowing freshets to pass over rather than impounding water above the crest level behind the dam. Also, under flooding conditions, a movable dam would permit open navigation of vessels and rafts.[13] Lock and Dam No. 11 was located above Zanesville, at Ellis, but below the earlier lock and dam at Symmes Creek, completed in 1838. This new location would ensure an adequate depth to the pool behind the Zanesville Lock and Dam. The army engineers had considerable experience with French movable dams on the Great Kanawha and Ohio Rivers, but in those cases, the Chanoine wicket dam was employed, and by 1929, fifty-two such structures were built on the Ohio River and a further ten on the Great Kanawha River. The first movable dams were erected and in operation before the Corps of Engineers acquired the Muskingum system in 1886, and it was not until 1910 that the Ellis Lock and Dam was opened for traffic. The Boulé movable dam, a variant on the Chanoine wicket system, featured panels that could be removed panel by panel (see figs. 2.6 and 2.7). These wickets were supported on trestles erected on five-foot centers with a depth of four feet over the fixed crest.

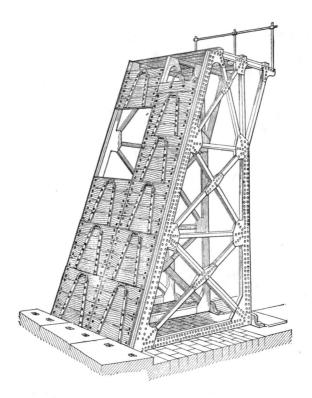

Locks

In order to provide for the passage of steamboats, the locks above Marietta were built approximately 228 feet long and 35.5 feet wide to accommodate vessels 162.72 feet long and 35 feet wide inside the lock chamber. This size was selected to represent the current and expected steamboat dimensions using the Muskingum. They were not large enough, though, to accommodate typical vessels plying the Ohio River. Lock and Dam No. 1 was sufficiently large (360 feet by 56 feet) to accommodate Ohio River boats, thus providing a so-called ice harbor for steamboats during the winter, as well as ensuring

Figure 2.6. The Boulé Dam raised Lock and Dam No. 11. This was the only movable dam on the Muskingum Navigation. The improvement was introduced by engineer Boulé at the Port L'Anglais near Paris in 1875. (*The Design and Construction of Dams* [1911], by Edward Wegman)

that Marietta would serve as a river port. Ice harbors were also established at Parkersburg at the mouth of the Little Kanawha River and at Point Pleasant (at the confluence of the Great Kanawha and Ohio Rivers).

All of the original sets of locks, including the first at Harmon across the river from Marietta, were constructed in the technology of the day, namely, ashlar masonry walls backed by rubble stonework. The bottom of the lock chamber used a timber deck. Since the sleepers and the deck planking were always under water, rotting of the wood was obviated. When the tide lock of the Alexandria Canal was excavated in 1983, all of the deck timbers in this 1844 lock were found to be sound.[14]

In a sense, lock walls are a form of dam or retaining wall but, unlike these structures, they must be designed to withstand earth and water pressure imposed behind the walls, as well as pressures developed when the lock chamber is full. In photographs of early locks, ashlar masonry can be clearly seen on the faces of the lock walls; but, hidden behind the smooth masonry, rubble stonework was utilized to achieve the masonry wall thickness needed to withstand the pressures. The backfill behind the wall was not usually compacted but allowed to settle over time (see fig. 2.5).[15]

Figure 2.7. The only French movable dam on the Muskingum Navigation. (C. W. Sutphin collection)

35

It should be noted that many of the early canal locks in America were crib walls filled with stone and the inside face of the chambers lined with planks. They are a near relative of the crib dam. None of these crib-wall locks were used on either the Ohio and Erie Canal or the Muskingum Navigation.[16]

Lock floors were usually timber planks spiked to a series of wooden sleepers that, in turn, were bolted to timber piles holding the floor down when the chamber was empty. With the high water table, there was an uplift pressure that would float the floors if they were not anchored down. Concrete was used only in later lock designs like Lock and Dam No. 11 on the Muskingum (see fig. 2.8).[17]

Lock Gates

With chambers thirty-six feet wide, the timber gates were unusually large and heavy compared to towpath canal gates with chambers fifteen feet wide. In the beginning, the gates were operated with geared drum winches and chains. Balance beams, traditionally used for miter gates, could not be used with such large gates, necessitating a mechanical means of opening and closing the gates (see fig. 2.9). According to Raber, Malone, and Parrott,

Figure 2.8. A view down the length of Lock No. 5 Muskingum River improvement. (C. W. Sutphin collection)

Failure of river wall allowing upper river gate to sag.
Lock 8 Muskingum River. July, 1932

two of these winches survive on guard gates at Beverly and Zanesville.[18] The most obvious means of attaching the chain would be at the top of the lock gate. Such a simple arrangement would tend to warp the gates and, in time, lead to ill-fitting closure of the miters. The answer was to attach the chain at the bottom of the gate and through a series of sheaves run up the side wall where the winch was mounted. While sound in conception, this arrangement tended to collect debris and required heavy maintenance. These early devices are quite rare; extant examples are found in Canada on the Rideau Canal and are also featured in the waterpower canal network at Lowell, Massachusetts.[19]

A more efficient drive for opening the lock gates is the rack and pinion, which consists of a long tooth-rail, or rack, pinned to the lock gate and engaged by a circular horizontal gear, the pinion, which is actuated by a crank. In addition to the Muskingum slackwater system, the nearest extant example is a rack and pinion on locks in the Little Kanawha.[20] The rack-and-pinion system for opening lock gates was extensively used on the Ohio and Muskingum Rivers.

Figure 2.9. A view of the Muskingum Lock. The leakage through the wood miter gate is a clear indication that gate repairs are needed. (C. W. Sutphin collection)

Gates

While a variety of designs have been used for lock gates, such as the roller gates used on the Ohio River or the drop gates employed on the Lehigh Canal, miter gates were the most popular and used exclusively on the Muskingum and the Ohio and Erie Canal. In the early days, gates were framed with vertical timbers, but for most ordinary later locks, horizontal timbers extending from the heel post to the miter toe were used. The toe ends were mitered so that the two gates closed in a vee shape to reduce leakage. The vertical heel post was shaped to fit the quoin in the lock-gate wall recess. White oak was the preferred wood for such gates, but yellow pine was also used. The heavy timber frames of the gate were sheathed with timber planks, creating a solid, watertight wall to resist water pressure.

Lock Chamber Valves

The common type of water valve used for medium- and low-head locks is the wicket, balanced, or butterfly valve—all of these terms are used. On towpath canals constructed before the Civil War, the filling and emptying of lock chambers was accomplished by pivoting wickets mounted in the lock gates and operated through gears and cranks at the top of the gates. While effective, this system was slow in moving water in and out of the lock chambers for large locks designed to accommodate steamboats. These wickets are constructed with a vertical shaft operated by a crank mounted on top of the gate. Traditionally, the wicket was cast iron with wrought-iron driveshafts and cranks (see fig. 2.9).[21]

The French "Fontaine" valve found favor in America, where it was named a cylinder or drum valve. This type of valve was installed after the Corps of Engineers assumed control of the Muskingum Waterway in 1886. A cylinder valve consists of a fixed outside cylinder with a movable vertical cylinder within the outer shell. The cylinder is raised and lowered with an iron stem attached to a ridged cone mounted on top of the cylinder. Raising the cylinder allows water to pass from conduit to lock chamber.[22] A pair of valves mounted on either side of the miter sill connected across the lock chamber with a culvert pierced by

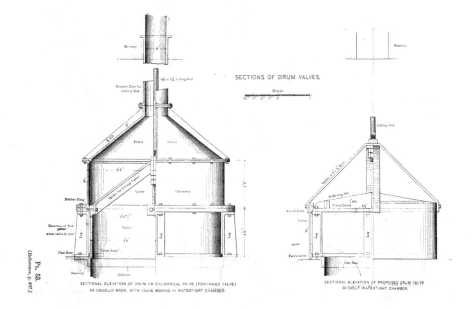

rectangular orifices below the miter sill at the head of the lock chamber. The valve, mounted in a conduit at the side of the lock, fills the chamber from a port or ports in the wall of the lock chamber. It provides a much quicker means of filling the lock than the installation of wickets in the lock gates.

The Corps of Engineers to the Rescue

During the first decade following the opening of the Muskingum waterway in 1841, considerable traffic developed, providing revenue for the waterway that was supplemented by receipts from the sale of waterpower at various locations. The improvement served as a catalyst in the industrial and domestic expansion of communities dotted along the course of the Muskingum River. Even though the railway did not arrive along the river until 1888, various railway companies thereafter built lines crisscrossing the watershed. This caused a steady decline of passenger and goods traffic on the waterway. As we have seen earlier, the decline also gripped the entire Ohio Canal system, prompting

Figure 2.10. Fontaine/drum valve. (*The Improvement of Rivers* [1913], by Benjamin Franklin Thomas and David Alexander Watt)

the Ohio General Assembly to lease the entire system at the beginning of the Civil War in 1861; this lasted until 1878. In an extraordinary arrangement, the lessees made no payment to the state but rather were allowed to retain all of the tolls collected and apparently received $100,000 from the state for operating costs. Without specific provision for maintenance, the lessees were not compelled to keep the waterway in good working order. Not surprisingly, it deteriorated. The situation became so bad that the lock and dam at Symmes Creek (formerly called Lock and Dam No. 1) was not even operable by 1877.[23]

Rescue was, however, on the way. The Federal River and Harbor Act of July 5, 1884, included funds for a survey of the Muskingum Waterway. The report was submitted on January 9, 1886.[24] The salient point of the report recommended that the waterway be transferred to the United States. The Ohio General Assembly acted quickly and on May 14, 1886, agreed to transfer all of the assets of the Muskingum River Improvement to the United States, effective July 1. The United States Congress in turn accepted the offer in the River and Harbor Act of August 5, 1886. The federal government felt that, to be viable,

> The provisions of this act, as far as they relate to the Muskingum River, shall not take effect, nor shall the money hereby appropriated be available, until the State of Ohio, noting by its duly authorized agent, turns over to the United States all property asked by the act of the General Assembly aforesaid, and all personal property belonging to the improvement aforesaid, and used in its care and improvement, and any balance of money appropriated by said state for the improvement of said river, and which is not expended on the fifteenth day of July, eighteen hundred and eighty-six.[25]

The consummation of the transfer was delayed until the general assembly acted on March 21, 1887. The United States took possession on April 7 of that year. The Muskingum, together with other tributaries of the Ohio, now formed part of a network of river navigations stretching nearly 1,000 miles from the confluence of the Monongahela and Allegheny Rivers in Pittsburgh to Cairo, where it joined the Mississippi River, including tributaries.

In the case of the Muskingum River, all of the dams and a number of the locks, except Lock and Dam No. 1 at Symmes Creek, were rebuilt. The skirting canal at Philo was abandoned and a new Lock No. 9 built at the end of the dam. To provide greater slackwater pool depths, the crests of five of the dams were increased as part of the rebuilding program.

Even before the United States' acquisition of the Muskingum River Improvement, the Corps of Engineers had been busy building an "ice harbor" at Marietta to shelter steamboats from the ravages of ice floes on the Ohio River during the winter. In the case of Marietta, the ice harbor involved building a new lock at Marietta at the other end of the dam and abandoning the original Harmon Lock. The new lock, which had an unusual hourglass shape with three pairs of gates, was begun in 1880 but was still under construction when the United States acquired the Muskingum River Improvement. Ohio River steamboats required more lock-chamber space than was provided in the locks and dams above Marietta, which were only 36 feet by 186 feet clear in the chamber and thus unable to accommodate the typical Ohio River steamboat. As a result, the number of steamboats that could shelter in the winter was greatly limited. Lock and Dam No. 1 (No. 11 in the old system) featured the so-called Monongahela-size lock chamber of 55 feet by 360 feet in the clear. Despite the need, the work dragged on until 1891.[26]

Symmes Lock and Dam had not functioned for many years, but by the end of the century there was renewed interest in re-establishing the connection between the Muskingum River Navigation and the Ohio and Erie Canal with a view of providing an expanded link between Lake Erie and the Ohio River. The replacement of the Symmes Creek Lock and Dam was a key to re-opening navigation from Dresden to the canal. In a typical approach, the Corps of Engineers undertook a survey of the upper reaches of the river, authorized by the River and Harbor Act of August 17, 1894. On the basis of the survey, the estimated cost for a new lock and dam at a location downstream from the original Lock and Dam No. 1 located at Ellis was $110,000. This new lock and dam would increase the navigable depth to six feet. A subsequent River and Harbor Act of 1902 provided the resources for a survey of the river from Zanesville to its headwaters at Coshocton. This report did not recommend improvement up to the confluence of the Walhonding and Tuscarawas Rivers at Coshocton but favored an improvement only as far as Dresden and then through the side cut to join the Ohio and Erie Canal.

The next River and Harbor Act, March 3, 1905, included approval of the $110,000 estimate for the lock and dam at Ellis—provided that not less than $200,000 be secured from non-federal sources—for improvement of the Ohio and Erie Canal from its junction with the side cut to Cleveland on Lake Erie. The conditions were met and the lock and dam constructed 2.5 miles below the old Lock and Dam No. 1. Being located farther downstream and with a higher crest on the dam, there was thought to be danger of flooding upstream. As mentioned earlier, the solution was to employ a French-style Boulé movable dam, the only such structure on the waterway.[27] Begun in 1907, the work was completed in 1910. While the navigation depth above Zanesville, linking with the Ohio and Erie Canal, provided a four-foot draft (the same depth as on the Ohio and Erie Canal), the later Ellis Lock and Dam increased the effective depth to six feet.

Earlier, the flood of 1847 had caused considerable damage to the crib dams and, in repairing the damage, the slope dams were rebuilt as step dams. Subsequently, all of the dams were converted to this step configuration as shown in figure 2.4. After the responsibility for the locks and dams was transferred to the Corps of Engineers in 1887, the step dams were reconstructed as slope dams, their original configuration.[28] Furthermore, from 1902 to 1922 concrete caps were added to each dam, except for No. 7 and No. 10, where only partial caps were installed. With the completion of repair work in 1922, the waterway was in better condition for navigation than it had been for decades. These improvements did not attract additional traffic. In fact, the gigantic flood of 1913 became a defining point in the history of the entire watershed, resulting in a deep concern for flood-control measures on the Muskingum watershed and elsewhere in the state of Ohio.

From 1824 to 1913 the steam packet boat ruled supreme on the Muskingum River, conveying passengers and a wide variety of goods and raw materials. Literally hundreds of steamboats operated on the river and ventured out onto the wider reaches of the Ohio River. Details of these steamboats form an appendix of Gamble's *Steamboats on the Muskingum* (see fig. 2.11).[29]

Following the great flood of 1913, barge and gasoline tows replaced the steam packet boats carrying coal, sand, gravel, petroleum products, and bulk cement. Canal traffic was confined to the last three slackwater pools to supply fuel to the coal-fired electricity-generating station at Philo. Although trade had diminished over the years, the decade of 1922–1931 shows a considerable tonnage of raw materials. The commerce was largely internal, with fewer than 10,000 tons leaving or entering the waterway at Marietta. Future projections at the depth of the Great Depression were based largely on the amount of coal, sand, and gravel that would be moved in the future.[30]

To improve the potential of the slackwater system, a nine-foot-deep channel with high-lift locks and movable crest dams was proposed. This is precisely the solution used on the Great Kanawha River. Beginning in 1933, four high-lift dams with German movable roller gates were constructed as part of the Public Works Program of the New Deal. Such a system for the Muskingum attracted little attention from federal New Deal agencies; rather, a comprehensive flood-control system on the Muskingum would become a model of interagency cooperation, with a focus on conservation.

With the completion of the nine-foot channel at the Belesville Locks and Dams, the slackwater pool on the Ohio River was raised to flood the Muskingum Marietta Lock No. 1. Thus, neither the lock nor the low dam was required. In 1955, the *W. P. Snyder* passed through Lock No. 1 to serve as a permanent exhibition of the steamboat museum at Campus Martius. She was the last steamboat to use the lock, which was removed in 1969.[31]

Its commercial traffic was diminishing, so the United States transferred the Muskingum slackwater system back to the state of Ohio on October 16, 1958.[32] With this legal action, the waterway once again became the property of the state of Ohio, as it was in the beginning. Under the aegis of the Ohio Department of Natural Resources, a program was initiated to repair the locks and dams for recreation purposes. Completed in 1967, this work permitted the slackwater system from Marietta to Zanesville to be opened for pleasure craft.

Other Slackwater Navigations

In an effort to reach the rich coal resources in Pennsylvania and West Virginia, a company was formed to improve the Monongahela River. In 1841 Locks and Dams No. 1 and No. 2 on the lower reaches of the river, a mile above the Smithfield Street Bridge, were completed. The new navigation

company, authorized by an act of the Pennsylvania Assembly in 1836, built a slackwater navigation from Pittsburgh south to the Virginia line. By 1856, the system was open for navigation to the Virginia line above Point Marion, Pennsylvania.[33] As in the case of the Muskingum Navigation, the United States acquired the slackwater system and upgraded it as late as the decades following World War II, the most recent work being the Grays Landing Lock and Dam, completed in the 1990s. The mining industry in the Monongahela Valley has decreased rapidly following passage of the Clean Air Act in 1970 because of the high sulfur content of the coal, and the new locks and dams in West Virginia and Pennsylvania have not been fully utilized since they were opened in recent decades.

The other river that forms the Ohio River at Pittsburgh, the Allegheny, was skirted by a towpath canal with locks as part of the Pennsylvania Main Line Canal, which opened in 1834. It was not until later, however, that locks and dams were erected on the river to provide a slackwater navigation in the lower reaches of the Allegheny.

Sensing rich coal and oil deposits upriver on the Little Kanawha River, a company formed in 1847 proceeded to construct four locks and dams to serve the Volcano oil field and timber interests upstream from Parkersburg, West

Figure 2.11. Notice the narrow navigation of the steam packet.

Virginia. With financial support from the United States government, a fifth lock was completed in 1891, called the Government Lock. This provided slackwater navigation to Creston, West Virginia. The Corps of Engineers in 1905 acquired the entire system. Surveys were made with a view to extending the system to Bulltown and the salt deposits associated with that area. It would have required ten additional locks and dams. The extension was never built.[34]

Moving from the Little to the Great Kanawha, a series of ten locks and dams were begun in 1873 and completed in 1898. These were French Chanoine movable dams, which provided open navigation during periods of favorable river stages and, in other times of the year, were raised to produce a series of slackwater pools.[35] Without question, the ten movable dams that created a slackwater system from the falls of the Kanawha to its confluence with the Ohio at Point Pleasant rank among the most successful projects ever completed by the Corps of Engineers. Opened for traffic in 1898, the slackwater system proved an economic success. By 1929, it became necessary to undertake major repairs on the oldest crib dams, but it was decided to abandon the entire system and replace it with four high-lift dams and twin 55-foot by 360-foot locks, the so-called Monongahela size. With increased coal traffic, a new 800-foot by 110-foot lock has recently been opened to

Fig. 2.12. An all-wooden coal barge.

navigation at Winfield, followed by a similarly sized lock at Marmet. By any evaluation, the canalization of the Great Kanawha River is an engineering and economic triumph.[36]

A much less successful project, but nevertheless of considerable interest, is the Big Sandy slackwater system, which used the French Poirée and Chanoine movable dams. Completed in 1897, the first Poirée needle dam and associated lock at Louisa marked the beginning of the slackwater system constructed by the Corps of Engineers. The needle dam at Louisa became the first such movable dam constructed by the Corps of Engineers, followed by four other needle dams. The well-tried Chanoine wicket dams were later installed on the weirs of the other dams. The main channel locks and dams on the Big Sandy were open in 1905, followed by Lock and Dam No. 1 on the Tug Fork and Lock and Dam No. 1 on the Louisa Fork in 1909. Although intended as a complete slackwater system, construction ended in 1914 and was never renewed. In fact, the system was abandoned and the movable dams dropped on the crest level to provide open navigation should it be required.[37]

The tributaries of the upper Ohio River presented both opportunities and challenges for the Corps of Engineers in their attempt to establish a comprehensive river navigation system. Conceived with great enthusiasm, many of these slackwater systems on tributaries of the Ohio did not undergo careful benefit-cost studies, but were nevertheless built. When unable to meet financial obligations, private companies were often acquired using political means; ownership was subsequently transferred to the United States with the Corps of Engineers designated to be responsible for operations.

Conclusions

While not an extreme case, the Muskingum River Improvement fits the pattern with its assets being transferred to the United States. The transfer of the system back to the state of Ohio is an example for the region and one hopes a harbinger of other waterways being transferred for recreational purposes only. Taken together, the Ohio and Erie Canal and the Muskingum River slackwater system represent the first attempt to tame the Muskingum watershed, not originally for flood control, but to provide a much-needed transportation network as part of the national Internal Improvements Movement, the hallmark of public works in the nineteenth century.

NOTES

1. Jay Mack Gamble, *Steamboats on the Muskingum* (Staten Island, NY: Steamboat Historical Society, 1971), 1.
2. Emory L. Kemp, *The Great Kanawha Navigation* (Pittsburgh, PA: University of Pittsburgh Press, 2000), 16.
3. John Kilbourne, *Public Documents Concerning the Ohio Canals* (Columbus, OH: I. N. Whiting, 1832), 44.
4. C. C. Huntington and C. P. McClelland, *History of the Ohio Canals: Their Construction, Cost, Use and Partial Abandonment* (Columbus, OH: Ohio State Archaeological and Historical Society, 1905), 158–160, 163–165.
5. Kilbourne, *Public Documents*, 298.
6. Kilbourne, *Public Documents*, 317–319.
7. Ohio Board of Canal Commissioners, *14th Annual Report of the Board of Canal Commissioners* (Columbus, OH: James Gardiner Printers, 1836), 10.
8. Kilbourne, *Public Documents*, 447–448. Charles Singer et al., *A History of Technology* (London: Oxford University Press, 1957), vol. 3, 440–444.
9. Kilbourne, *Public Documents*, 317, 359.
10. Kilbourne, *Public Documents*, 404.
11. Edward Wegman, *The Design and Construction of Dams* (New York: John Wiley and Sons, 1900), 140–145.
12. B. F. Thomas and D. A. Watt, *The Improvement of Rivers* (New York: John Wiley and Sons, 1913), part 2, 530, 608–616; Wegman, *Design and Construction of Dams*, 162–164.
13. Thomas S. Hahn and Emory L. Kemp, *The Alexandria Canal: Its History and Preservation*, Monograph Series No. 1, Institute for the History of Technology and Industrial Archaeology, Morgantown, WV: West Virginia University Press, 1993.
14. Thomas and Watt, *Improvement of Rivers*, plate 53; Michael S. Raber, Patrick M. Malone, and Charles Parrott, *Muskingum River Lock and Dam Study*, Revised Draft Report, Unpublished (Columbus, OH: Wolpert Consultants, 1991), 18. Raber, Malone, and Parrott, *Muskingum River Lock and Dam Study*, 18.
15. Thomas and Watt, *Improvement of Rivers*, 392–408.
16. Thomas and Watt, *Improvement of Rivers*, 410.
17. Raber, Malone, and Parrott, *Muskingum River Lock and Dam Study*, 12–14.
18. Raber, Malone, and Parrott, *Muskingum River Lock and Dam Study*, 19–20.

19. Larry Sypolt and Emory L. Kemp, "The Little Kanawha Navigation," *Canal History and Technology Proceedings*, vol. X (Easton, PA: Canal History and Technology Press, 1991), 66–68.
20. Thomas and Watt, *Improvement of Rivers*, 493–496.
21. Thomas and Watt, *Improvement of Rivers*, 497–498. War Department, Corps of Engineers, *Report on the Muskingum River, Ohio, Covering Navigation, Flood Control, Power Development, and Irrigation* (Huntington, WV: United States Engineering Office, Huntington, WV, December 10, 1932), 23; RG 77, entry 111, file 7249 Bulkies (Preliminary Examinations–Muskingum River), 23, NARA–Philadelphia.
22. U.S. Army Corps of Engineers, *Annual Report* (Washington: GPO, 1886), part 2, 1552.
23. *Report on the Muskingum River*, 23A.
24. *Report on the Muskingum River*, 24.
25. *Report on the Muskingum River*, 26.
26. *Report on the Muskingum River*, 29–31.
27. Thomas and Watt, *Improvements of Rivers*, 241.
28. Wegman, *Design and Construction of Dams*, 150.
29. Gamble, *Steamboats on the Muskingum*, 130–141.
30. *Report on the Muskingum River*, 30–31.
31. Gamble, *Steamboats on the Muskingum*, 114.
32. Gamble, *Steamboats on the Muskingum*, 115–116.
33. *Monongahela Navigation Co.* (Pittsburgh, PA: Bakewell and Marthens, 1873), 617. Sypolt and Kemp, "Little Kanawha Navigation," 30.
34. *The Great Kanawha Navigation*, 42–82.
35. *The Great Kanawha Navigation*, 249–252.
36. *The Great Kanawha Navigation*, 252
37. Leland R. Johnson, *Men, Mountains and Rivers: An Illustrated History of the Huntington District, U.S. Army Corps of Engineers* (Washington: GPO, 1977), 108–113; Thomas and Watt, *Improvement of Rivers*, 563–592, 688.

CHAPTER THREE

The Politics of Flood Control

T he first attempt to utilize the water resources of the Muskingum watershed was for navigational purposes, with no attempt to control floods.
The Ohio and Erie Canal and then the Muskingum Improvement provided,
for the first time, a transportation network that opened up hitherto inac-
cessible regions for both agriculture and industrial development. The rise of
communities along the Muskingum and its tributaries can be attributed in
large measure to these two waterways.

In the case of the Ohio and Erie Canal, reservoirs were built at the Portage
Summit near Akron on the divide between the Licking River and the Scioto
watershed to supply water to the highest elevations of the canal. These reservoirs were not designed to, nor were they able, to control floods but rather
ensured that the canal would have sufficient water during the boating season.
In a similar matter, the locks and dams on the Kanawha River Improvement
established a series of slackwater pools in stair-step fashion. No attempt was
made to control floods. Any freshet was passed over the crest of the dams and
into the Ohio River.

Floods could not be avoided; as we have seen, the climate and geography
produce notable floods throughout the upper Ohio Valley—including the
Muskingum watershed.

Even before the American Revolution, destructive flooding occurred at
Fort Pitt at the close of the French and Indian War in 1763. Floods were a
constant threat, but the lure of rich bottomland for crops, waterpower sites,
and transportation in the Muskingum watershed lured settlers. Devastating
floods for which there are records occurred February 1832, April 1860, February 1884, March 1898, March 1907, March 1910, March 1913, August
1935, January 1937, and April 1940.[1]

These "acts of God" were accepted by the public; it was felt little could be done to control such vast forces of nature. From the 1820s, the U.S. Army Corps of Engineers had responsibility for navigation on the inland waters of the nation. In part, this responsibility was founded on national defense in much the same way that Eisenhower's interstate defense highway system was promoted in the national interest following World War II. The mission was single-minded and clear: the Corps of Engineers would undertake public works to improve navigation on the inland rivers. Flood control, power generation, irrigation were not considered to be part of their mandate.

Following a series of floods on the lower Mississippi below the confluence with the Ohio River at Cairo, Illinois, in the autumn of 1851, Stephen Harriman Long of the topographical engineers was ordered to meet with his colleague Captain Andrew A. Humphreys in Philadelphia, in response to a congressional mandate to examine the nation's principal waterways with a view toward controlling flooding for navigational purposes. In addition to Humphreys and Long, included on the engineer board was Charles Ellet Jr. (1810–1862) who, years earlier in 1849, had completed the famous Wheeling Suspension Bridge across the Ohio River. It was during the construction of this bridge that Ellet turned his attention to river hydraulics. Ever the individualist, he did not wish to serve on a panel but rather to make his own recommendations, which he did in a report to the Corps of Engineers in January 1852. Subsequently, he published this work under the title *The Mississippi and Ohio Rivers.*[2] The Ellet report recommended the traditional approach of using higher and stronger levees, supplemented with "cut offs" such as a controlling channel to the Atchafalaya watershed. The centerpiece of his report, however, was the construction of various reservoirs on the tributaries and headwaters of the Mississippi, Missouri, Red, and Ohio Rivers.

Humphreys and his assistant, Henry L. Abbot, spent more than a decade (interrupted by the Civil War) to prepare their report. It became the classic nineteenth-century American study of river hydrology. It was respected not only in America but also abroad. Their detailed report established the "levees-only" doctrine of flood control that guided the Corps of Engineers until 1927. Not only was the report widely accepted by the Corps of Engineers, but the authors rose to become generals and, in addition, Humphreys assumed the position of chief of engineers the year the report was published.[3]

Ellet's work was discredited by Milnor Roberts, a distinguished engineer, contending that the construction of a plethora of high dams was not feasible. In terms of cost and engineering technology of the day he may well have been right. His opinion was underscored by none other than Colonel William Emery Merrill, father of the Ohio slackwater system.[4]

Following the establishment of the levees-only policy, Congress authorized improvements of the lower Mississippi levee system under the guise of improving navigation. Steadfast opposition claimed that flood-control measures were of local benefit only, not in the national interest, and thus unconstitutional.[5]

For six decades, from 1866 to 1926, the Corps of Engineers stuck by its levees-only position in the face of heavy flooding in back-to-back floods in 1883 and 1884 that dealt a severe blow to the upper Ohio Valley region. The St. Valentine's Day flood inundated communities along the Muskingum and its tributaries, with fourteen feet of water in the case of Marietta, Ohio. Federal, state, and local rescue missions were mounted. The lack of flood-control measures meant the role of the federal government was reduced to disaster relief only. The first covert federal projects for flood control, under the rubric of navigation, resulted from these two disastrous floods. Earlier, in 1874, tentative steps were undertaken by Congress to improve the levee system on the lower Mississippi River. Congress also authorized levees around selected Ohio River towns.

Zanesville provides a vignette of the effect of flooding in communities along the Muskingum River and its tributaries. During the course of construction of the Muskingum slackwater system, a levee for bank protection was erected along the riverfront, and repairs were undertaken by citizens of the town following damage caused by the flood of 1860. Subsequent flooding in 1869 and the major flood of 1884 damaged the levees. On these occasions repairs were undertaken, in the latter case, by the state of Ohio. Following acquisition of the slackwater system by the Corps of Engineers in 1887, many citizens believed that the Corps should be responsible for flood damage if not, in fact, for flood control. Their idea was tested in the flood of 1898. In a pork-barrel tactic, a grant of $6,000 was added to the bill for harbor improvements in Cleveland. Over strong opposition by the army engineers that the work was inappropriate to their mission, the chief of engineers acquiesced to the will of Congress. In similar circumstances the wall failed again in 1910, and again, under protest, the Corps of Engineers were required to repair it in response to the orders of Congress.[6]

Like in the great flood at Fort Pitt in 1763, Pittsburgh was inundated by a devastating flood in 1907. Much of what is now the Golden Triangle was underwater. Built at the confluence of the Allegheny and the Monongahela Rivers, the center of Pittsburgh was quite vulnerable to inundation. It was estimated that in the 1907 flood, $6.5 million in damage occurred. In a very real way, floods punctuated the politics of flood control. An elective group of individuals, firms, and agencies bonded together to promote flood-control measures to protect Pittsburgh. As the first in the nation, except for work on the lower Mississippi, it pioneered the promotion of flood-control legislation. The Pittsburgh Flood Control Commission issued a comprehensive report on flood control, championing the idea of reservoirs, levees or flood walls, and conservation measures such as afforestation.[7]

Eventually a notable series of dams was built on the headwaters of the Allegheny and Monongahela Rivers, which provided the protection sought by the Pittsburgh Flood Commission. These were owned and operated by the Corps of Engineers. Just a year after the Pittsburgh Flood Control Commission report, a great flood inundated vast areas of the Ohio Valley, including the Muskingum watershed. Torrential rain moving from the west across Ohio, which caused great damage in the Miami and Scioto Valleys, hit the Muskingum watershed after previous rains had already saturated the ground. The heavy rain began on March 23, 1913, and persisted for five days. Fortunately, the storm ended over the western part of the Muskingum basin, reducing by thirty-five percent the level of flooding on the Tuscarawas River. Nevertheless, eleven lives were lost, and eighteen bridges on the Muskingum slackwater and fifty-one on the tributaries were destroyed. In Zanesville alone, 3,400 buildings were flooded, including 157 swept away or shifted on their foundations. Utilities were disrupted; roads and railways were cut off. It was the worst flood ever recorded in the area. It was estimated at the time that damages exceeded $20 million.[8]

Flood Heights of Record in the Basin[9]

Streams and Locality	Elevation		Extreme Flood Heights								
	Low Water	Damaging Flood Stage	1832 Feb.	1860 April	1884 Feb.	1898 March	1904 Jan.	1907 March	1910 March	1913 March	1913 July
Muskingum Rivers:											
Marietta	583.2	602.0	617.2	612.4	621.3	615.6	610.7	619.5	609.9	627.1	595.8
McConnelsville	640.5	660.0	670.0	667.0	669.0	669.5	661.2	660.0	667.7	684.0	662.8
Zanesville [1]	672.9	690.7	700.2	698.8	699.7	702.5	692.9	697.6	698.4	717.4	688.7
Coshocton	735.3	750.0	...	...	758.9	760.9	752.8	755.3	755.4	766.4	745.0
Tuscarawas River:											
Canal Dover	859.0	866.7	...	...	...	879.0	...	...	...	885.6	...
Zoarsville	866.2	875.0	...	...	...	886.0	...	...	...	891.2	...
Massillon	914.0	924.0	...	...	...	934.0	...	...	...	940.0	...
Warwick	943.1	951.0	...	...	...	959.0	...	...	...	965.0	...
Reeds Run	...	...	...	...	...	878.0	...	...	...	882.2	...
Sandyville	905.0	913.0	...	...	...	914.0	...	...	...	920.0	...
Walhonding River:											
Walhonding	815.3	823.0	...	...	...	836.0	...	...	...	884.0	..
Killbuck	795.0	803.0	...	...	...	806.0	...	...	...	810.7	...
Millersburg	806.5	814.0	...	...	...	817.0	...	...	...	821.5	...
Brinkhaven	860.0	869.0	...	...	...	874.0	...	...	...	861.7	...
Loudenville	933.2	941.0	...	...	...	946.5	...	...	...	954.6	...
Mount Vernon	974.1	981.0	...	...	...	986.5	...	...	...	988.0	...
Mansfield	1,133.0	1,139.0	...	...	...	1,140.5	1,149.2	...	...	1,142.7	...
Wooster	848.0	855.0	...	...	...	857.0	...	...	...	860.4	...
Lexington	1,162.0	1,170.0	...	...	...	1,191.0	...	...	...	1,200.0	...
Wills Creek:											
Cambridge	770.0	785.0	...	...	...	...	...	796.0 [2]	...	794.2	

1—The zero of Zanesville Gage is 665.9 feet elevation; 2—Local storm

The table provides dates in which to compare principal floods beginning in 1832. Zanesville has been flooded seventeen times, for an average of once in 5.88 years. The great flood of 1913 can be categorized as a 100-year-frequency flood. The record and costly floods of 1907 and 1913 in the Ohio Valley led directly to federal action in the matter of flood control. First, a House Committee on Flood Control was empaneled in 1916, leading directly to the first Flood Control Act of 1917. Opposition in the form of proponents of river and harbor projects, which were regularly sponsored by Congress, felt that this would dilute funding available for their projects. The timing was fortuitous, since the Mississippi was once again in flood in 1917. The bill passed Congress with little opposition.[10]

The act appropriated federal funds directly to flood control, for the first time with no justification in terms of benefits to navigation. Lingering concerns about local benefits vis-à-vis national interests resulted in the idea of local participation in flood-control projects sponsored by the federal government. Equally important, it authorized the Corps of Engineers to undertake flood-control studies but still linked such studies to navigation, hydroelectric power generation, and other uses in watershed areas—presumably with conservation in mind. The federal government was now committed to more than its historic single concern for navigation.[11]

Following World War I, Congress became increasingly interested in developing hydroelectric resources and enacted the Water Power Act of 1920, which, amongst other provisions, established the Federal Power Commission. The power commission later had a significant role to play in New Deal public works projects such as the rebuilding of the Great Kanawha Navigation with high-lift dams suitable for the installation of power-generating equipment.[12] As a result of the 1920 act, the Corps of Engineers launched a comprehensive waterpower study. The results, published in 1926, are commonly referred to as the 308 and 309 documents. The report stated:

> There are evidently two principal purposes for which investigations of this nature would be useful, either for the preparation of plans for improvement to be undertaken by the federal government alone or in connection with private enterprise or to secure adequate data to insure that waterway developments by private enterprise would fit into the general plan for the full utilization of the water resources of a stream. This department is now charged with examinations and surveys for navigation and flood control improvements, and with the construction of such projects as are authorized by Congress. In both cases of investigations, departments must by law give consideration to the development of potential water power.[13]

The basis of the study was the investigation of 180 rivers, and it was perhaps the most comprehensive study ever produced by any agency of the federal government.

No sooner had the 308/309 Report been published than the 1927 Mississippi flood occurred, perhaps the greatest flood disaster to strike the inland rivers of America. The death toll stood at 250, with 700,000 homeless. At the time, the damage was estimated at $230 million in direct costs and another $200 million in indirect costs that were difficult to assess. From the perspective of flood control, this flood swept away in its path the "levees-only" policy cherished by the Corps of Engineers since the publication of the Humphreys and Abbot report. This policy could no longer be defended; it was discredited as levees failed in the lower Mississippi below Cairo and were shown to be quite inadequate. In a futile attempt to exonerate the Corps of Engineers, General Edgar Jadwin, chief of engineers, defended his earlier plan for flood control, which in the aftermath of the 1927 flood was judged to be inadequate by both experts and the public.[14]

Despite the 1927 flood and the Flood Control Act of 1928, the Corps of Engineers was still wedded to the "navigation only" policy. It was not until the Flood Control Act of 1936 that the Corps of Engineers was unshackled from this policy to take on a more comprehensive approach to water management that included flood control. More recently, the Corps of Engineers has been environmentally conscious and a steward of not only the natural environment, but also of cultural resources.

To understand the history of the Muskingum reservoir system, it is necessary to turn attention to the aftermath of the 1913 flood and move west to the Great Miami River Valley. This populated valley was particularly hard hit with the death of nearly 500 people and property damage in excess of $300 million attributed to this one storm alone. The damage was spread across the state, including the Muskingum watershed and the city of Zanesville. The public response was not only to clean up the mess but to address the problem of flood control in the Great Miami Valley. To this end, a citizens' group was formed and raised $2,130,000. At the behest of Dayton City Engineer Gaylord Cummin, they recommended former colleague Arthur E. Morgan and his firm, Morgan Engineering Company of Memphis, Tennessee, to undertake a study of the Miami River Valley with a view toward controlling future flooding. This investigation proved to be the beginning of a long and fruitful relationship, with an engineering approach to flood control. Morgan was involved with the reservoir system of the Muskingum watershed and was the first chairman of the Tennessee Valley Authority. A person who did not attend college, he nonetheless became president of Antioch College. In addition to other achievements in a long and distinguished career, his firm, which he founded in 1909, was engaged on the Miami flood-control project.[15]

In a farsighted and fortuitous move, the Ohio legislature secured a constitutional amendment to provide protection of natural resources and the establishment of conservation districts, effective January 1, 1913. Morgan presented a model bill to establish conservancy districts, labeled as House Bill No. 19, which became effective in February 1914. Such districts could pursue the following projects:

A. Preventing floods

B. Regulating stream channels by realignment, widening, and/or deepening

C. Reclaiming or filling wet and overflow lands

D. Providing for irrigation

E. Regulating stream flows

F. Developing watercourses [16]

A historic event occurred on January 28, 1915, when the Miami Conservancy District was founded. The plan, prepared by Morgan, was adopted on May 10, 1916, and guided the work that began in 1918 and was substantially completed in 1923.[17]

As we have seen, during the first half of the twentieth century there was considerable interest in hydroelectric power generation. The 308/309 Report emphasized development of hydroelectric power. In a day of increased concern for the conservation of natural resources and the reduction of air pollution, hydroelectric power appeared to be highly desirable. In the case of flood-control dams, this was shown to be clearly not feasible. To operate flood-control reservoirs, it is necessary to lower the pool level during the winter and early spring to provide maximum storage for floodwaters. The pool level is raised in the late spring and early summer to ensure an adequate amount of water for low-water augmentation for navigation and water-pollution control in late summer. For power generation, on the other hand, the ideal situation is to have the maximum pool level to provide the greatest "head" for power generation. While power-generation equipment was installed at the three high-lift, roller-gated dams on the Great Kanawha River in the 1930s, power generation was not part of the Miami or later Muskingum flood-control projects.

In fact, on both of these systems—the Miami and the Muskingum—retarding basins (or dry dams) were featured and became a new technology for water-resource management in the eastern part of the United States. All five dams on the Great Miami River scheme were of this type, while only key installations featured dry dams on the fourteen reservoirs of the Muskingum system. In a dry-dam design, an earth or concrete dam was constructed in the usual manner, with overflow spillways to prevent any possibility of overtopping the dam in an exceptional flood situation. If such an overflow event occurs, the dam would be quickly destroyed. The special features in these dams were concrete conduits designed with cross sections that allowed a controlled amount of water, able to be safely carried by the watercourse below, to be let through the dam during freshets. The remaining water was impounded behind the dam awaiting the time when the floodwaters could be safely released downstream. It is important to note that there are no gates to control the flow, and thus the structure is "self-acting." Under normal conditions, the river simply flows through the conduits, unimpeded by the dam, allowing the water in the reservoir area to be utilized for agricultural purposes.

The work on the Great Miami system was begun under wartime conditions in 1918. With labor and material shortages, the district undertook the work as a general contractor, and hence no contracts were signed. The project, consisting of five dry dams and associated flood-control structures, was completed in 1923. In the construction of the five earth dams, the hydraulic-fill method, first developed in the western United States, was employed almost exclusively. Apparently this was an early application of the system in the eastern part of the United States. It is an ingenious method whereby very fine material is floated into the core, forming an impervious barrier against the movement of water through the dam, while the coarse material forms embankments on either side of the core. Hydraulic fills were not employed on the Muskingum project.

With the Great Miami flood-control scheme complete and functioning, it served as an example for those living along the flood-prone Muskingum River and its tributaries. Since the 1913 flood, communities had sought local means of protection against destructive floods. After much study, it became apparent that the problem was bigger than any of the communities and beyond their financial resources. Thus, in 1927 the Zanesville Chamber of Commerce took the lead and canvassed other communities about following the example of the Miami Conservancy District in a comprehensive approach to flood mitigation. The initiative proved to be quite timely, since the nation was suffering from the effects of the 1927 floods. As mentioned earlier, it was the flooding of the lower Mississippi that discredited the cherished "levees-only" policy of the Corps of Engineers, thus providing engineers with other possibilities for flood control. In the beginning of his involvement with water-resource management, Morgan was a proponent of the "levees-only" policy, but he quickly abandoned it in favor of the dry-dam scheme for the Great Miami River.

The stage was now set for the founding of the Muskingum Water Conservancy District.[18] In an enlightened action, the Zanesville Chamber of Commerce selected Bryce C. Browning to head the committee on flood control. He guided the committee through a labyrinth of actions and requirements to emerge victorious with the founding of the Conservancy.

Bryce Browning, who more than any other person is credited with the creation and success of the Muskingum Watershed Conservancy District in the flood-control project, served as its secretary from 1933 until his retirement in 1965.

While active in both church and community organizations, his greatest contribution to the people of the Muskingum Valley, undoubtedly, was his advocacy of flood control. From 1927 (when he assumed the role of secretary of the Chamber of Commerce) until 1965, he championed flood-control measures. In a recent newspaper article Chuck Martin wrote that Browning was tireless in his effort, overseeing the many facets of the effort and making numerous trips to Washington, D.C., both to lobby for the project and to meet with officials as the great effort proceeded.[19]

During his long tenure as secretary of the MWCD, numerous awards and citations were bestowed upon Browning for his work in conservation. The successful partnership of the MWCD and the U.S. Army Corps of Engineers rests in no small part on the leadership he displayed and his determination to see the project through to a successful conclusion.

One of the first acts in January 1928 was to engage Morgan and his Dayton-Morgan engineering firm to prepare a study of possible flood-control measures that could be undertaken to curb flooding throughout the watershed area. The report, presented in June 1928, was comprehensive in nature. It concluded that partial protection could be provided at Zanesville, but complete protection would require the mobilization of forces throughout the counties represented in the watershed. An important insight concerned the necessary role of the federal government if "complete" protection was to be achieved. The report called for the establishment of a conservancy district similar to the Miami Conservancy District, satisfying the provisions of the Conservancy Act. Of the nine separate plans presented in the report, six would offer protection for a flood of the magnitude of that of 1913, whereas the remaining three would offer only partial protection but at a much lower cost. Influenced by the success of the Miami scheme, the report considered retarding reservoirs (dry reservoirs), together with levees, as possible approaches to mitigating the ravages of floods. Thinking beyond just the Zanesville situation, the report investigated twenty-three possible reservoir sites in the watershed area.[20] At the same time, the cities of Dover and New Philadelphia on the Tuscarawas River, a principal branch of the Muskingum, were organizing to combat flooding much the same as Zanesville. A similar group was formed at Massillon to support not only flood mitigation, but also improved navigation, their idea being the development of a barge canal connecting the Ohio River and Lake Erie.

Following an area-wide meeting on April 29, 1930, interested groups met at New Philadelphia and formed the Muskingum-Tuscarawas Improvement Association. During July 1930, this association received a $10,000 grant to undertake a detailed study of the river system with a view of recommending measures about future flood control in the valley. Not surprisingly, the study was undertaken by the Dayton-Morgan Engineering Company. For the first time the study encompassed the entire watershed. Published on June 24, 1931, the report said in part:

The flood problem is complicated and difficult of solution because of the large size of the drainage area and the wide separation of the communities that need protection. These conditions make it difficult, if not impossible, for some communities to get adequate protection by local plans, and the only recourse is to cooperate with other communities in a general plan which would provide such protection. Hence, no one flood-control method would be adequate. The effect of forestation on severe floods is very slight. While check dams may perform a useful function in soil and water conservation and controlling local floods, . . . they can not alone, however, be relied on as a major factor in controlling large floods . . . Their effect on large floods in the Muskingum Valley, like that of 1913, would be very slight. Moreover their cost per acre foot of storage would be about 10 times that of large reservoirs.[21]

The use of reservoirs, both conventional with impounding lakes and also dry reservoirs, formed the centerpiece of the report, but levees and flood walls in selected communities were found to be desirable together with channel improvements. Despite the proponents of a barge canal or indeed a slackwater system as far as Dover, the report did not favor such an investment, feeling that its cost would override its benefits. This document proved to be crucial in securing large-scale federal support as part of the national New Deal public-works program. On June 3, 1933, the Conservancy Court, under the authority of the Ohio Conservancy Act, created the Muskingum Watershed Conservancy District. Its mission included not only flood control, but water conservation, the control of soil erosion, and the general development of water uses in the entire watershed. The Conservancy County Court consisted of judges of the Court of Common Pleas from each of the counties in the district and was established by the Conservancy Act of Ohio.

Shortly after its formation in August 1933, the Muskingum Watershed Conservancy District applied to the Federal Emergency Administration of Public Works to become part of the New Deal public-works program. The request sought federal funds to implement the reservoir plan and associated flood-prevention measures. Eager to see large-scale projects to provide work for the unemployed, the Public Works Administration noted that the engineering plan was already in hand, making an early construction start possible, and readily granted $22,090,000 to the Corps of Engineers. The Corps of Engineers was to be responsible for the engineering and construction of fourteen reservoirs and the relocation of all public utilities, roads, and railways impacted by the reservoirs. The grant, dated December 1933, led to an agreement on March 29, 1934, with the MWCD. Survey and foundation investigations of approximately 150 possible sites, of which fourteen were selected, were undertaken almost immediately. The detailed plan submitted by the Corps of Engineers received approval on November 19, 1934. At the same time, contract drawings and specifications were advertised and contracts let by the end of the year of 1934. This cooperative arrangement involved the Corps of Engineers, the Ohio State Assembly, and the Ohio State Highway Department, as well as the MWCD. The state highway department, in cooperation with the Federal Bureau of Public Roads, undertook the relocation of highways in the reservoir areas at an initial cost of $9,919,106.

The Conservancy District assumed responsibility for acquiring land, rights of way, and flood easements for the project at an estimated cost of $9,050,610. Titles were acquired or easements obtained for more than 7,500 properties, in addition to 700 rights-of-way for utilities companies. This cost was covered in part by the state assembly granting $2,000,000, leaving the remaining $7,000,000 to be paid by land owners who would benefit directly from flood-control mitigation.

Several supplementary grants were obtained by the MWCD, one in 1937 for $1,600,000 and a second for $4,500,600 resulting from provisions of the Flood Control Act of 1938. These were in recognition of the increased cost for land transactions incurred by the MWCD. As a result, the assessments leveled by the MWCD were entirely eliminated.

Earlier, $3,500,000 had supplemented the original construction amount provided to the Corps of Engineers.[22] With finances secured, the work proceeded rapidly. Each of the partners fulfilled its assigned role, enabling the project to be completed in just four years. It was indeed a triumph for the new cooperative approach to a large-scale public works.

Before continuing the story of the construction of the fourteen dams, it is well to consider the engineering techniques brought to bear on both earth and concrete dams and the hydrology undergirding the designs. In many ways the engineering represented a new chapter in a comprehensive approach to water-resources management.

NOTES

1. Secretary of War, *Muskingum River and Its Tributaries, Ohio,* H.R. Doc. 251, 78th Cong., 1st sess., 1943, 5.

2. Charles Ellet Jr., *The Mississippi and Ohio Rivers* (Philadelphia: Lippincott Grambo and Co., 1853), 387.

3. Martin Reuss, "Andrew A. Humphreys and the Development of Hydraulic Engineering; Politics and Technology in the Army Corps of Engineers 1850–1950," *Technology and Culture* 26 (Jan 1985), 1–33.

4. Leland R. Johnson, *Men, Mountains and Rivers* (Washington: GPO, 1977), 142.

5. Johnson, *Men, Mountains and Rivers,* 142.

6. Johnson, *Men, Mountains and Rivers,* 145–146.

7. Pittsburgh Flood Control Commission, Report (Pittsburgh: 1912); Joseph L. Arnold, *The Evolution of the 1936 Flood Control Act* (Fort Belvoir, VA: Office of History, U.S. Army Corps of Engineers, 1988), 14.

8. *Report on the Muskingum River, Ohio* (Huntington, WV: War Dept. U.S. Engineer Office, 1932), 42–43, RG77, series 111, entry 7249 Bulkies, Archives II.

9. *Report of the Ohio Valley Flood Board* (Columbus, OH, House Document No. 1792, 12 Sept 1914), table 49.

10. *Report on the Muskingum River,* 44–45. Arnold, *Evolution of the 1936 Flood Control Act,* 13–15.

11. Emory L. Kemp, *The Great Kanawha Navigation* (Pittsburgh: Pittsburgh University Press, 2000), 223–247; Hal Jenkins, *A Valley Renewed* (Kent, OH: Kent State Univ. Press, 1976), 94.

12. Arnold, *Evolution of the 1936 Flood Control Act,* 18–19.

13. Hal Jenkins, *A Valley Renewed,* 2; Arthur E. Morgan, *The Miami Conservancy District* (New York: McGraw-Hill Book Co., 1951), 11–62; Arthur E. Morgan, *Report of the Chief Engineer, Submitting a Plan for the Protection of the District from Flood Damage,* 3 vols. (Dayton, OH: 1916), 17–23; Kemp, *The Great Kanawha Navigation,* 223–247.

14. Jenkins, *A Valley Renewed,* 6–9.

15. Jenkins, *A Valley Renewed,* 64.

16. Jenkins, *A Valley Renewed,* 9.

17. Jenkins, *A Valley Renewed,* 9; Morgan, *The Miami Conservancy District,* 16–19.

18. Jenkins, *A Valley Renewed,* 13–15.

19. Chuck Martin, "Banker Bryce Browning Made His Mark in Conservancy," *Times-Recorder* newspaper, Zanesville, OH, 13 Nov 1999, 1–4.

20. Jenkins, *A Valley Renewed*, 45.

21. Jenkins, *A Valley Renewed*, 68–71.

22. "$3,500,000 More Allotted Flood Plan," unpublished scrapbook, Muskingum Watershed Conservancy District, New Philadelphia, OH.

CHAPTER FOUR

Built of Earth and Concrete

B efore presenting details of the Muskingum reservoirs, it is appropriate to consider the "state of the art" with regard to both earth and concrete hydraulic structures. Earth levees and dams have a long tradition based upon experience. This empirical approach established "rules of thumb" without benefit of theory. For instance, it was held that earth dams required the use of impervious materials throughout the cross-section of the dam if the structure were to be watertight. In the twentieth century, numerous successful dams have been built of a wide variety of materials with modern theories to show how to control the seepage.

Theoretical means for calculating earth pressures date from the work of Coulomb (1736–1806), who made use of the phenomena of soil cohesion and internal friction. Later, in the Victorian era, the theoretical basis of earth pressures was further refined by C. V. Poncelet (1835), W. J. M. Rankine (1857), Karl Culmann (1866), Wilhelm Ritter (1879), and others.[1] As a discipline, however, one must credit Karl von Terzaghi. In 1925 he published a seminal paper entitled *Erdbaumechanik*.[2] Largely forgotten in the field was the presentation and later publication of pioneering work on earth dams by Justin. Joel D. Justin, a hydroelectric consulting engineer, published his sixty-two-page paper in the ASCE Transactions in 1924. This was accompanied by a forty-nine-page set of discussions by leading civil engineers.[3] Justin enhanced his work with publication in 1927 of the *Hydro-Electric Handbook*, consisting of more than one thousand pages ranging from the construction of earth dams to the engineering aspects of hydro-electric power generation.[4] Adding to his published work was a 1932 book devoted entirely to earth dams.[5]

By the beginning of the Muskingum reservoir project, about 1933, Justin had laid the engineering foundations for earth-dam design and construction, whether for flood control, irrigation, power generation, water supply for communities, or even navigation systems. The Muskingum reservoir system, under the aegis of the Corps of Engineers, followed what was then the latest technology in both earth embankment and concrete gravity design and construction, including essential structures such as gated conduits to pass controlled floods through dams as well as spillways to ensure that the dam crest would never be overtopped, while at the same time ensuring that erosion below the toe of the dam was controlled.

In an effort to clarify the technology, a discussion of the various steps undertaken in the design of both earthen-embankment and concrete-gravity dams is helpful. The next chapter will illustrate the practical engineering application of the Justin approach to the subject.

Flood Data

Detailed and accurate flood-control data are essential in any attempt to control freshets by means such as dams, dikes, levees, or flood walls. Factors influencing inundation are presented in chapter 1. With accurate data stretching over a long period of time, a flood-control program can be established based upon a statistical approach. For example, a system could be designed to cope with the expected 500-year return flood. In any case, a benefit-cost study would reveal for a given amount how much protection could be obtained. If full protection cannot be secured, how much damage can be permitted? This obviously is a function of the return cycle of floods in a given watershed. Much more damage would be sustained if heavy, damaging floods occurred every ten years rather than every 500 years.

Flood data for the Muskingum watershed is known in terms of the dates of damaging floods and the extent of inundation, but rainfall, runoff, and detailed flood data before the twentieth century are largely unknown. As a result, a statistical approach was not used, but rather the reservoir system for the entire watershed was based upon the 1913 flood, thought to be the most severe to hit the area since European settlement (more details are shown in the table in chapter 3, page 53). In order to provide an adequate factor of safety, the design flood resulted from the basic 1913 flood data plus a forty percent increase. At the time, this was thought to be quite adequate and, indeed, it has provided security for the entire watershed to this day. With more sophisticated prediction models and a longer period for studying inundations in the area, there is now a concern that the system does not provide for a predicted 500-year return flood. Studies are underway to evaluate each of the dams with a view to increasing reservoir capacity by selectively increasing crest heights, in conjunction with other flood-control structures such as flood walls.

Initial Site Surveys

To build such a system, a detailed survey of the entire watershed is undertaken to select whether to use dams, associated reservoirs, and other hydraulic structures such as flood walls. Surveys typically begin not in the field but in the office, using topographical maps and aerial surveys to determine the lay of the land throughout the watershed. At the same time, property surveys are made with a view to producing comprehensive property maps. In the case of the Muskingum watershed, it was the Watershed Conservancy (MWCD) that was responsible for the purchase of lands or for obtaining easements in connection with not only the area for the reservoirs itself, but also the relocation of utilities and transportation networks such as roads and railways. From both the office and fieldwork data, storage capacities can be determined for each potential site. Additional information can be obtained on stream-flow characteristics from soundings in the riverbed and by the use of flood meters. Conservation studies for both agriculture and forestry should be undertaken at the same time so that this information can be integrated into the final selection of reservoir sites. In the case of the Muskingum, preliminary surveys of the watershed identified 150 potential sites of which fourteen were selected for the erection of dams and associated reservoirs.

Subsurface Exploration

Once a site is selected, subsurface investigations need to be undertaken to determine the permeability and bearing capacity of the materials encountered on the site, and sources for erecting earth embankments also need to be identified. Equally important, comprehensive material sampling needs to be done and analyzed for possible use in the embankments. It might appear that a rock foundation is the most sought-after site condition. Core borings, however, need to be made to determine if faults, cleavages, or veins are present that could provide open channels for water to penetrate through the dam.[6] Techniques used at the time included test pits, test trenches, and core borings utilizing several boring techniques. Knowledge of subsurface conditions is also essential for the design of concrete gravity dams as well as embankments built of earth, since the bearing capacity of the material is essential information for both types of dams. Whatever subsurface investigation technique is used, careful consideration needs to be given to the number and location of testing locations. To ensure that the best site can be selected and that any fault lines can be detected, it is certainly false economy to restrict the number of bore holes or test pits located in a potential reservoir area. In the case of the Muskingum Conversancy District, elaborate subsurface investigations were undertaken at each of the fourteen dam sites.

Field Soils Laboratory

In order to analyze the samples obtained in the field, laboratory results need to be obtained for both the rolled-layer method or the hydraulic method of building an earth embankment and also for bearing capacity and shear values for concrete gravity dams. By the time the Muskingum project was underway in the 1930s, a wide range of soil types had been successfully used in a variety of locations. In an effort to expedite the field investigations so that construction of dams could be undertaken quickly, a soils field laboratory was set up in Zanesville by the Corps of Engineers, who were responsible for the construction aspects of the project.[7] Regardless of materials used, the upstream portion of the dam should be built in a series of rolled layers and be nearly impervious, whereas the downstream face can be constructed of gravel or even small boulders. This is especially true of the toe of the downstream part of the dam.

In the case of the hydraulic-fill earth dams, such as the five constructed in the Miami Conservancy District, it is essential that the fill material contain enough fines such as sands or clay to be deposited in the core of the dam to restrict the flow of water through the cross-section. Clays composed of

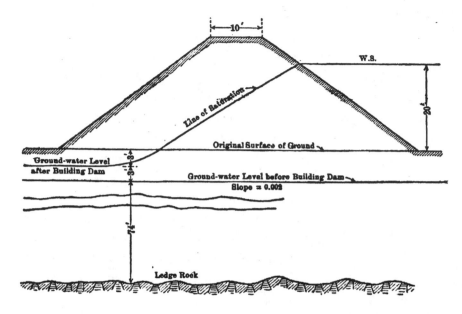

particle sizes less than 0.005 mm, the smallest grains in clay, are so small that they are in fact colloidal, which gives clay its cohesiveness and characteristic imperviousness. Thus clays, in combination with sand or silt, can produce a nearly impervious core for both hydraulic-fill and roller earth dams.

Line of Saturation

The line of saturation and hydraulic gradient are important concepts used in the design of earth dams. Regardless of the material employed in constructing an earth dam, water will migrate through the structure. The highest line of flow is called the saturation line (see fig. 4.1).[8] If the saturation line issues above the downstream toe of the dam, serious consequences may ensue. If, for instance, the downstream portion of the dam is composed of fine sand or silt, erosion may result from the development of channels in such fine materials, causing the dam to fail. If, on the other hand, the toe portion of the dam is composed of clay, small "boils and land slides" may result, leading to possible failure. For downstream faces built with coarse gravel or even stones, the flow through the dam can be controlled, and such a dam may be quite

Figure 4.1. Effect of earth dam on ground-water level showing the line of saturation.

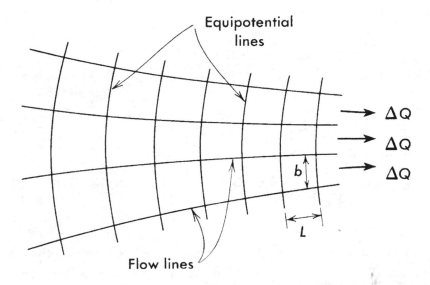

Equipotential lines

Flow lines

safe. The general rule, however, is to design the cross section of such a shape and with a suitable gradation of materials that the line of saturation is kept below the toe. A core wall of concrete, masonry, or clay will have little effect on the line of saturation if it does not cut off flow under the dam, according to Justin in his discussion of factors influencing seepage in dams.[9]

The old rule of thumb that earth dams should be constructed of impervious material is clearly not correct or even necessary; rather, the hydraulic design focuses on the seepage of water through the dam from the upstream to the downstream elevation. Any seepage analysis must include the foundation or the natural material upon which the dam rests. Recognizing that seepage occurs in earth dams, several assumptions allow engineers to analyze the flow patterns and determine a satisfactory design for a given site and embankment materials.

Since the flow in the absence of piping or fissures will be quite slow and devoid of turbulence, the well-known Darcy formula for laminar flow is used: $Q = K i A$, where Q is the seepage in cubic feet per second, i is the hydraulic gradient in feet per foot, and K is the permeability factor of the material in feet per second and can be determined in laboratory investigations of typical soil samples. The hydraulic gradient, as applied to embankments, may be

Figure 4.2. Analytical flow net. (Kemp collection)

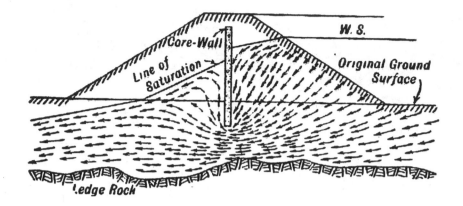

defined as the gradient of a line joining the highest points to which water would rise in a series of pipes placed in the cross section. This line is generally, but not necessarily, coincident with the line of saturation. For the modern watershed project, piezometers have been installed in each of the dams so that the water levels can be continuously monitored. The potential head, measured in feet, is the sum of the elevation at the pressure head at any point. Hydraulic gradient is thus the linear rate of drop through the cross section of the dam. In a typical hydraulic study, it is assumed that the seepage can be determined using a two-dimensional cross section.

At the time of the design of the Muskingum watershed reservoirs, the usual approach to determining seepage was a graphical method employing a "flow net" (see fig. 4.2). This solution consists of a series of trials and adjustments of a flow net until a satisfactory solution is obtained. The flow net consists of a series of equipotential lines intersecting a set of flow lines at right angles. The flow net is composed of a series of approximately square figures. The flow can, therefore, be considered as a series of tubes labeled Delta Q, yielding the flow per foot of dam.[10] On the basis of the seepage analysis, cutoff walls, rock toes, or even downstream stilling basins, spillways, and conduits can be determined. Flow of water through an earth dam is illustrated in figure 4.3.

Figure 4.3. Flow of water through an earth dam.

An alternate approach is to use scale-model studies in a laboratory. To obtain reproducible results, materials such as rubber, plaster of Paris, and plastics have been used. It is essential that the deformation characteristics of these materials as measured by "Poisson's ratio" be employed. Equally important, it is not just a matter of making the model and the prototype geometrically similar, but the model must produce correct results for the dynamic forces involved. Thus, such a study requires dynamic similitude as well as simple geometric similitude. Laboratory studies were available in the case of the MWCD to determine the hydraulic characteristics of conduits flowing full or partially full, which aids in the design of the structures necessary to pass water through the dam (see fig. 4.4).[11] Overflow spillways are essential to prevent unusually severe flooding from overtopping the dam. The shape is in effect a large rectangular weir with a nappe that conforms to the natural shape of the free-flowing trajectory issuing across the top of the weir. This shape provides the maximum discharge flow while at the same time

Figure 4.4. Beach City Dam showing the conduit outlet below the earth dam. (Photo by Kemp)

preventing cavitation on the concrete surface below the weir. Cavitation can cause severe erosion of the spillway surface. At the foot of the spillway, the stream is returned to a horizontal position and discharges into a distilling basin to dissipate energy and avoid stream erosion below the dam. The first use of this spillway configuration was by John Jervis in the Croton Dam, New York, 1836–1846. Spillway design of this "ogee" shape has changed little since that time. Such a design lends itself to model structures to determine the shape and capacity for an individual situation.

Shaft or morning-glory spillways provide an alternative means and are in considerable contrast with the conventional ogee shape (see fig. 4.5). Elegant in appearance, these are really a variant of the closed conduit with the intake functioning as a weir. Such a spillway is featured at the Muskingum watershed Pleasant Hill Dam.[12]

Figure 4.5. Elegant morning-glory spillway design of Charles Mill Dam. (*Applied Hydraulics in Engineering*, by Henry M. Morris [The Ronald Press Co.: New York, 1963]: 205)

Slope Stability

Both up- and downstream slopes of an earth dam are required to be stable under all loading conditions existing with impounded water, i.e., elevations varying from zero to a maximum height. The required slope will, of course, depend upon the materials. Unlike the usual case, the angle of repose of the material on the upstream face should be the underwater angle, which usually will be much flatter than the ordinary angle of repose. It is prudent to make the slope even flatter than the test results that determine the underwater angle of repose. In practice, this means no slopes more than half the angle of repose. The downstream face must safely meet two criteria. First, the slope must not exceed the dry angle of repose and, in fact, a more gentle slope is recommended. Second, the slope must be adjusted so that the saturation line intersects the base safely within the toe section.

Even the most satisfactory design of a rolled-layer dam will be compromised unless the material is compacted to obtain maximum density to avoid leaks and landslides. The construction specifications should reflect the means of obtaining the maximum density with the most appropriate moisture content and compacting equipment. Once the embankment is completed, great care must be taken to protect up- and downstream slopes. If, perchance, the downstream slope is composed of sizable rock and boulders, no ground cover is needed. In those reservoirs with a long fetch of wind, wave action on the upstream face of the dam may be quite damaging. Perhaps the most successful means of coping with wave action is the use of riprap, either random or hand placed. Concrete placed in blocks or even continuous slabs is a useful alternative.

The two principle means of constructing an earthen dam are the hydraulic-fill method used exclusively on the first Miami Conservancy District dams, and the more common rolled-layer method used throughout the Muskingum Watershed (see fig. 4.6). The hydraulic-fill method has been described above. Dams, levees, and canal and railway embankments have been built simply by dumping materials from trestles. Although widely used for railway embankments, this method is to be avoided in the case of hydraulic structures. The only exception is the semi-hydraulic method, in which the material is deposited

on or near the site then placed in the embankment by sluicing with water. As a general rule, layers should not exceed eight inches in thickness, compacted to a prescribed density. Both the rollers and moisture content are important factors in achieving a successful dam that avoids creating channels for the free flow of water through the dam or even water penetrating through the foundation material. From the contractor's point of view, it should be noted that the least amount of compacting effort occurs with the optimum moisture content. It is, therefore, critical that moisture values for the material be carefully controlled as the layers are placed in the dam to achieve the desired results at the least expenditure of compacting effort (see figs. 4.7 and 4.8).

From antiquity, the most common dam is the masonry dam and, later, the concrete gravity dam. Such a dam resists water impounded behind it by the sheer force of gravity. Water, wave, and ice forces act horizontally, tending to overturn the dam about its toe. The weight of the concrete acting vertically resists this type of failure (see figs. 4.9 and 4.10).

Figure 4.6. A view along the Beach City Dam showing the huge size of the earth dam when completed.

Since the hydraulic forces act horizontally, they also tend to cause sliding of the structure downstream. The sliding, or shear, acting at the bottom of the dam is resisted by friction developed by the weight of the structure pressing against the potential sliding plain. At any horizontal section, the plain concrete must resist the tendency to fail by shear. The foundation conditions must also be carefully investigated to avoid a shear failure in the foundation material, even though the dam itself may be adequate. Such a study would determine if the foundation was overstressed (see figs. 4.11, 4.12, and 4.13).[13]

Figure 4.7. Concrete spillway under construction at Leesville Dam, 1937. The concrete forming is for the spillway. (US Army Corps of Engineers)

Figure 4.8. In an effort to employ as many men as possible, traditional hand and mule techniques were utilized as shown of the Glendening Conduit Tunnel. (Kemp collection) Figure 4.9. A special view of the Leesville-rolled earth embankment and form work for the concrete spillway. (US Army Corps of Engineers)

Figure 4.10. An additional view of the Charles Mill embankment and spillway. All but the spillway was completed in 1936. Details of the spillway are shown in the foreground. (US Army Corps of Engineers)

Figure 4.11. General view of the concrete gravity dam at Dover. (Photo by Kemp)

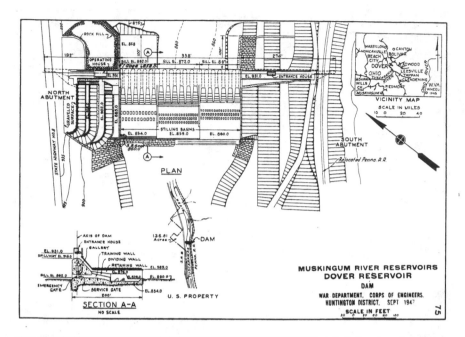

Figure 4.12. Plan of Dover Dam showing salient information of construction detail. (US Army Corps of Engineers)

Figure 4.13. Following an engineering investigation by the board of consultants, the Tappan Lake excavation for the spillway was greatly enlarged. (C. W. Sutphin collection)

NOTES

1. Hans Straub, *A History of Civil Engineering* (Cambridge, MA: The M.I.T. Press, 1964), 229.
2. Hale Sutherland and Raymond C. Reese, *Introduction to Reinforced Concrete Design* (New York: John Wiley & Sons, 1950), 147.
3. Joel D. Justin, "The Design of Earth Dams," *Transactions of the American Society of Civil Engineers*, vol. 87, 1924, 1–141.
4. *Engineers*, vol. 87, 1924, 1–141.
5. William P. Creager and Joel D. Justin, *Hydroelectric Handbook* (New York: John Wiley & Sons, 1950), 1–148.
6. William P. Creager, Joel D. Justin, and Julian Hinds, *Engineering for Dams Vol. 3* (New York: John Wiley & Sons, 1945).
7. Joel D. Justin, *Earth Dam Projects* (New York: John Wiley & Sons, 1932), 69.
8. Justin, *Earth Dam Projects*, 89.
9. Justin, *Earth Dam Projects*, 119.
10. Justin, *Earth Dam Projects*, 125.
11. Henry M. Morris, *Applied Hydraulics in Engineering* (New York: The Ronald Press Co., 1963), 180–181.
12. Morris, *Applied Hydraulics*, 204.
13. Morris, *Applied Hydraulics*, 171–178.

CHAPTER FIVE

Modern Mound Builders

A fter years of discussion and planning, the official agreement between the Muskingum Watershed Conservancy District and the United States of America was enacted on October 8, 1934, amended on April 15, 1935, and again on June 5, 1935. The agreement, entitled the "Official Plan for the Muskingum Watershed Conservancy District," is a comprehensive document spelling out the rights and responsibilities of all the partners involved, including the U.S. Army Corps of Engineers and the Muskingum Watershed Conservancy District as well as the Ohio Department of Highways. Other agencies such as the Civilian Conservation Corps and the Soil Conservation Service were also to participate in this grand scheme.[1] The Corps of Engineers assumed the responsibility for construction of the reservoirs and other flood-control structures, while the MWCD would acquire land and easements connected with each of the reservoirs and assume responsibility for building recreational structures and other matters. The Soil Conservation Services is now called the Natural Resources Conservation Service.

In this chapter, the emphasis is on the engineering and construction of the fourteen dams and other hydraulic structures incorporated in the official plan. The plan presented details of the history of the project; a statement of the flooding problem throughout the watershed; details of flood-control reservoirs including location of structures, the relocation of structures, railroads, highways, and public utilities; and the operational and conservation features associated with the plan. A description of the dams and reservoirs forms an important part of the document on the watershed. In all, the document exceeded 140 pages.[2] A diagrammatic sketch showing the location and reservoir capacity of each reservoir is included (see figure 5.1).

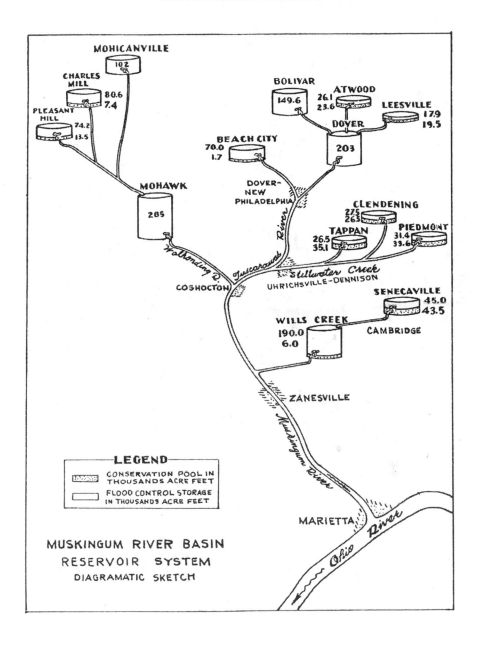

Figure 5.1. In an effort to show the location and reservoir capacity of each reservoir, a diagrammatic sketch is included.

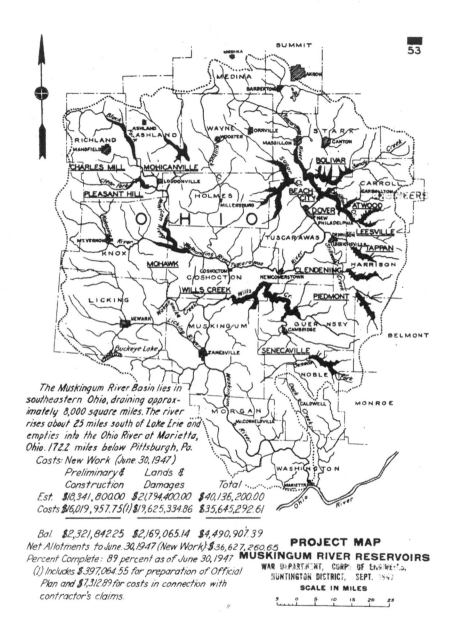

The Muskingum River Basin lies in southeastern Ohio, draining approx- imately 8,000 square miles. The river rises about 25 miles south of Lake Erie and empties into the Ohio River at Marietta, Ohio. 172.2 miles below Pittsburgh, Pa.

Costs: New Work (June. 30, 1947)

	Preliminary & Construction	Lands & Damages	Total
Est.	$18,341,800.00	$21,794,400.00	$40,136,200.00
Costs:	$16,019,957.75(1)	$19,625,334.86	$35,645,292.61
Bal.	$2,321,842.25	$2,169,065.14	$4,490,907.39

Net Allotments to June. 30,1947 (New Work) $36,627,260.65
Percent Complete: 89 percent as of June 30,1947
(1) Includes $.397,064.55 for preparation of Official Plan and $7,312.89 for costs in connection with contractor's claims.

PROJECT MAP
MUSKINGUM RIVER RESERVOIRS
WAR DEPARTMENT, CORPS OF ENGINEERS,
HUNTINGTON DISTRICT, SEPT. 1947
SCALE IN MILES

Figure 5.2. In addition to the diagrammatic sketch, a project map shows not only the dam sites, but also the extent of each reservoir.

According to the approved plan, the fourteen reservoirs were designed to reduce flood damage to a minimum with the funds available. Later, an additional seven projects were considered to enhance the system. These included three additional reservoirs and four local protection projects. The entire system was built on the basis of the "official plan flood" assuming ten inches of rainfall and a runoff of seven inches over a three-day period. Where it was not possible to provide both recreational and flood stages, the capacity of a reservoir was devoted to flood control. Thus, unlike the early Miami system of five "dry" reservoir dams, the Muskingum Watershed System included both "dry" and traditional "wet" reservoirs. It was assumed that the official plan flood was thirty-six percent greater than the 1913 flood, which was the most serious on record (see figure 5.2).

The system can be considered in a series of three groups, associated with the Walhonding, Tuscarawas, and Wills Creek drainage basins. The reservoirs

Figure 5.3. Building the spillway on Mohawk Dam. (US Army Corps of Engineers)

in the Walhonding basins consisted of the Mohawk, Pleasant Hill, Charles Mill, and Mohicanville, with the Mohawk Reservoir being the key to flood control in that basin (see figure 5.3 and 5.4). The Mohicanville Reservoir was to consist of a 500-acre lake, but it would be shallow and hard to maintain and thus was eliminated despite strong local pressure to the contrary. Both the Mohicanville and the Mohawk Reservoirs were for flood-water storage only and, hence, designated as "dry" reservoirs.

In a similar way, the Dover Reservoir is the key to the Tuscarawas River system. The Bolivar, Atwood, and Leesville Reservoirs are located upstream from the Dover Dam, while Beach City, Tappan, Clendening, and Piedmont are downstream. Although Dover was intended to have a small conservation or recreation pool because of heavy stream pollution, it was later operated as a "dry" reservoir along with the Bolivar Reservoir.[3]

Wills Creek is controlled by two reservoirs, each with conservation pools forming permanent lakes used for recreation and conservation purposes. The

Figure 5.4. Beach City spillway in operation. (Kemp collection)

Senecaville Reservoir is located on the upstream on the Seneca Fork, while the Wills Creek reservoir is downstream on Wills Creek itself.

Unlike the Miami Conservancy dams, which were built using the hydraulic-fill method mentioned above, the Muskingum dams were all built of rolled earth placed in thin layers that were compacted to specified densities. The sole exception is the Dover Dam, constructed entirely as a mass concrete gravity dam.

In contrast to the open conduit system on the Miami dams, each of the Muskingum dams has controlled outlets for releasing impounded water through the dam and discharging it into watercourses below the dam. The gated outlets were designed with the capacity to pass sufficient water, even if one of the gates did not function properly and could not be opened. During freshets, the volume of water released was controlled by the capacity of the downstream channel.

To avert a major catastrophe in the case of a 500-year flood, ungated spillways were installed below the crest of the dam to carry away any excess water and

Figure 5.5. Beach City Dam outlet conduit. (Kemp collection)

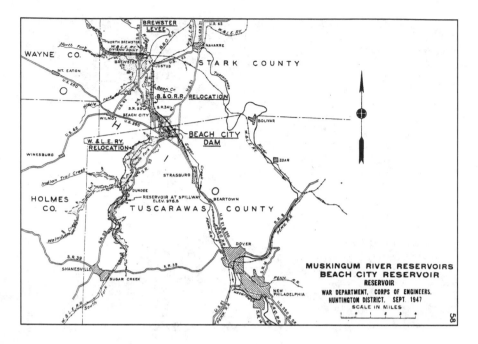

prevent overtopping of the dam, which, in the case of an earth-filled dam, would result in the embankment being washed away in part or even in its entirety.

In order to minimize nationwide unemployment at the earliest possible date in the depths of the Great Depression, leaders of the New Deal embarked on a number of large-scale public works designed to involve as many men as possible and to put them to work as soon as possible (see figs. 5.5, 5.6, and 5.7). Thus, the Muskingum and other New Deal projects were part of what today would be called "fast-track" projects.

Appointed as district engineer for a new Corps of Engineers District, Major J. D. Arthur established his headquarters in Zanesville, complete with a soil-mechanics laboratory. It was one of the first field laboratories in a large public-works project in the nation. The work of the laboratory was pivotal in the economic and engineering success of the entire project. As work progressed, several branch offices were also established. In all, more than 400 people were employed.

In the case of the MWCD, the operations were carried out by a board of directors chaired by W. O. Littick of Zanesville with T. J. Haley of Dover and

Figure 5.6. Beach City location map. (Project Maps, US Army Corps of Engineers, 1947)

Newton Mansfield of Ashland as committee members and Brice Browning serving as secretary. The MWCD maintained its own engineering staff and legal counsel, with C. C. Chambers serving as the chief engineer. In addition, judges from courts of common pleas from each of the counties involved in the project formed, in accordance with state law, the Conservancy Court:

> Simultaneous with the federal grant of funds for aiding in financing the work, the War Department established a district engineer's office in Zanesville. This organization now employing more than four hundred technical men and workers, is located with administration offices in the First National Bank building, engineering division on the fifth floor of the Masonic temple and with laboratory, garage, and storeroom in the Ellis garage building on North Fifth Street. Major J.D. Arthur, Jr., District Engineer, is in charge of operations.
>
> Major J.D. Arthur, Jr., District Engineer, U.S. Engineers, heads the local organization. Captain E.N. Chisolm, Jr., is chief of the operating division. Captain Albert C. Leiber, Jr., joined the staff last week. Lieutenant, F.S. Tandy is executive officer, T.T. Knappen is chief engineer.

Figure 5.7. Operating house for Glendening Dam. This structure releases water through the dam by means of the conduit at the base of the dam. (Photo by Kemp)

The conservancy district was created under the laws of the State of Ohio and the territory covered by the district is practically the entire 8,088 square miles of drainage area of the Muskingum river and its tributaries and the 288 square miles of Duck Creek drainage area, which total area is one firth of the State of Ohio. All or part of 16 counties are included in the district.

Operations of the conservancy district are carried on by a board of directors of which W.O Littick, Zanesville, was chairman with T.J. Haley, Dover and Newton Mansfield, Ashland co-members and Bryce C. Browning, Zanesville, secretary. The district maintains its own counsel and engineering staff and the common pleas judge from each of the counties in the district makes up the conservancy court. C.C. Chambers is chief engineer for the district.[4]

After the award of funds to construct the fourteen dams on December 29, 1933, other than reported surveys of the watershed area the public learned little of the engineering work underway. It was rather like a swan sailing serenely on a pond with little evidence of exertion, while under the surface it was paddling vigorously with its feet. In like manner, the engineers in the Zanesville district office spent the entire year of 1934 paddling vigorously as a prelude to construction. In addition to fieldwork necessary to determine the location of the dams (which involved soil sampling as well as topographical mapping to determine not only the elevation and cross section at each dam, but also extensive studies of the capacity of the reservoirs for the so-called official plan flood), engineering design was underway in the district office involving both the hydraulic and structural engineering design "tailor-made" for each site. Following the analysis and design of each dam, bid documents were issued, consisting of specifications for each aspect of the project together with detailed drawings of every aspect of the construction.

As an example, the hydraulic and structural engineering analysis revealed that the calculations were begun on the Beach City Dam on November 26 and completed on December 23, at the same time that the board of consultants was meeting.[5] The comprehensive design began with site selection as a prelude to the hydraulic design and actual construction of the flood-control reservoir. The official plan flood determined the outlet requirements for the gated concrete conduits. The discharge rate was carefully calculated and controlled so that the downstream channel would not be overburdened, causing excessive bank erosion.

With a five-foot freeboard below the crest of the dam, the spillway had to pass any excessive floodwaters above the plan-flood level. From the soil mechanics point of view, comprehensive site data formed the basis of the rolled-earth embankment. The foundation design was based on the pioneering work of Karl von Terzaghi. According to section VII of the design:

> The embankment will be constructed by rolled-fill methods. The central portion of the main embankment will be of imperious material, while the outer portion will be pervious. The dike will largely be impervious. A berm 10 feet wide will be provided at flow line to prevent washing of the slopes. The slopes will be made flatter at the base of the embankment for additional stability.[6]

All aspects of the embankment design were considered in great detail, both to make sure that nothing was overlooked that might compromise safety and to provide detailed requirements for contract purposes. Even such minor details as trash racks in the section of the outlet structure were considered. Since gated conduits represent the "Achilles' heel" of the whole design, the control gates would normally be operated by commercial electric power, but in case of a power failure at a critical time, a gasoline-powered generator was installed. As a further precaution, if both of the above failed, the gates could be manually operated. The discharge end of the design called for a concrete baffle system placed in the stilling basin to dissipate energy in an effort to reduce to a minimum any downstream channel erosion.

The location of the spillway constituted an important design consideration. The selection involved an assessment of the rock foundation for the structural concrete spillway and, equally important, the effect downstream of spillway discharge, if it ever occurred. The shape of the spillway was also carefully considered, resulting in an "ogee" shape.

The design specified low-heat cement for the concrete structures to avoid shrinkage cracks and a buildup of excessive heat in the large lifts of concrete for the conduit and the spillway. The construction sequence for the earthwork and concrete structures was carefully considered and specified, not leaving such a schedule to the discretion of the contractor.

The 143-page design document represents the state of the art in embankment design during the 1930s, incorporating as it did the use of soil mechanics and a foundation laboratory, current engineering, understanding of seepage, slope stability, and foundation-bearing capacity, plus many other important design considerations. The hydraulic design is equally impressive for reservoir, conduit, and spillways. Similar design documents were published for all of the other dams, levees, dikes, and relocation of utilities. It is an impressive example of the engineering aspects of large and complex hydraulic structures.[7] In a particularly notable development, the U.S. Engineers Soil Laboratory undertook a model study. Robert R. Philippe, the author of the report on a significant model study of the Mohawk Dam, describes the purpose and scope of the study as: "The main purpose of the study described in this report is to obtain a figure for the probable seepage at the proposed Mohawk Dam."[8] The study was made for the rolled-fill section shown on the official plan. The figure reported is for flood-stage conditions with headwaters at 890 feet and tailwater at 815 feet. A secondary purpose was to investigate the possibility of "piping" at dangerous locations near the dam and to investigate certain modifications of the design. The study was conducted by means of models constructed in the model tank (i.e., flume) in the soil laboratory and was supplemented by independent calculations by the writer.[9]

Built in a large flume, cross sections of three locations in the dam were studied. Soil information was gleaned on-site from undisturbed samples tested in the soils laboratory and from boring log data. In addition, Philippe inspected the site including test pits and by a more general study of the geological formations from gravel pits and terraces in the valley. The boring log locations of his report were dated August 28, 1934, whereas the first report was filed almost a year later in April 1935.

In an effort to determine theoretical values for seepage to compare with the model study, Philippe modified the "Terzaghi formula" to fit the site conditions. Comparisons of the theoretical to the model results showed good correlation, well within the accuracy of the field data, demonstrating that the model study approach was valid and could be utilized at other sites. Modifying the Terzaghi formula and securing model study results represents a noteworthy pioneering effort on the part of the Zanesville Engineer Office.

During the flurry of design activity during December 1934 at the Zanesville office, the Board of Consulting Engineers met to review the engineering designs with special reference to criteria for the design of spillways. In order to ensure that none of the dams would be overtopped even in the most severe floods, two criteria were developed:

Criterion 1

Start at the conservation pool level. That is the permanent pool level of the reservoir.

Develop hydrographic inflow volumes.

Duration of runoff three days.

Peak based on envelope of Miami basin March 1913 rates.

Outlets assumed to be closed.

Freeboard of 5 feet below the top of the dam.

Criterion 2

Start at the spillway crest elevation.

Determine hydrographic inflow volumes.

Duration of runoff three days.

Peak flow based on an envelope of Miami basin March 1913 flood rates.

Outlets assumed to be closed.

Freeboard of 2 feet below the top of the dam.[10]

For each reservoir, criteria 1 and 2 were to be applied. The spillway design was based on whichever criterion was most severe. Thus, the requirements fixed the minimum spillway dimensions needed as well as the spillway channels.

Still concerned with the capacity of each dam to discharge water under maximum flood conditions, the board at its January 1935 meeting turned its attention to outlet capacities.[11] After protracted discussion during which no decisions were made, the board advised that the outlets be increased as follows on the basis of volume in inches for a three-day flood. Atwood increased from 8.3 inches to 9 inches, Bolivar from 7.9 inches to 8.5 inches, Freeboard from 8.6 inches to 10 inches, and Leesville from 8.7 inches to 10 inches. When considering Wills Creek, the MWCD engineers felt that the outlet capacities for this dam should be reconsidered. Mr. Knappen, in charge of the designs for the Corps of Engineers, pointed out that the computations were very thorough and, furthermore, the contract for building the Wills Creek

Dam was already let, making it quite difficult to make radical changes in the contracts. To resolve this question, Mr. Harkness of the group representing the district office was instructed to prepare immediately runoff, outlet, and spillway requirements on the basis of a watershed of 300 square miles with a rapid runoff and 544 square miles for a less rapid runoff. While the emergency study was underway, William H. McAlpine of the Office of the Chief of Engineers felt the conduit should be increased by one-third. During this period he was, perhaps, the leading civilian engineer in the Corps, and his opinion was not to be taken lightly. Estimated at $125,000 to increase the capacity by one-half, his suggestion represented a twenty-four percent increase in this budgeted item. This involved increasing the six seven-inch by fifteen-inch gates to six eight-inch by seventeen-inch gates, which would increase the capacity from 23,000 cfs to 30,700 cfs. The larger gates were based upon a design for the Mohawk Dam. It was decided to increase the capacity of Wills Creek by enlarging the conduits to carry the full discharge capacity of the six seven-inch by fifteen-inch gate openings. Furthermore, the board decided to eliminate trash racks entirely from the Wills Creek, Beach City, and Mohawk Dams, which featured large conduit control gates.[12]

While the Beach City analysis serves as a case study for all of the rolled-earth embankment dams in the Muskingum Watershed system, the Dover Dam is the sole exception, being a concrete gravity structure. The analysis, however, followed the same outline as the Beach City report (see figs. 5.8 and 5.9). The selection of the site and type of dam comprised the first section, followed by the hydraulic design considerations, including reservoir capacity, conduit, and spillway design capacities using the same approach as on other dams. The essential difference is the design of the concrete structure:

> The dam as finally designed will be of massive concrete, straight gravity type, and was selected as the best suited and most economical because of the restrictions imposed by the narrow box like canyon, high flood discharges, and the proximity of good rock foundation. The central portion of the dam is ogee in shape to form the spillway section through the base of which are located the outlet works at three elevations. The stilling basin consists of a stepped apron with training walls between the conduits and a system of baffle piers immediately below in order to create a hydraulic jump before leaving the apron and passing into the natural river channel.[13]

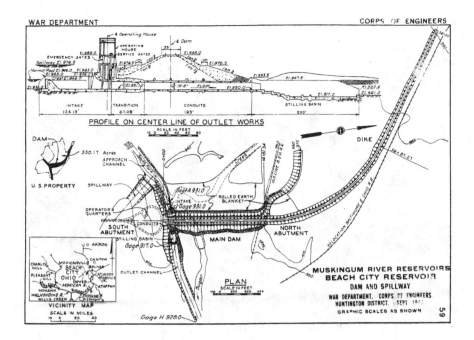

The foundation consists of approximately seven feet of sand and gravel overlying shale, sandstone, limestone, and coal. In designing the concrete structure and determining the foundation bearing capacity, the principal concern was sliding and uplift caused on the horizontal position of the bedding planes and, most importantly, the presence of shale, which would offer weak resistance to both shear and sliding. These beds are characterized as containing many water-bearing fractures. After studying the option, it was decided to use the lower Mercer limestone beds to support the dam and to provide a shear cutoff wall carried to the sandy silt shale at the head of the dam.

In order to ensure the dam's success and safety, the plan featured extensive provisions against uplift and the possible "piping" through grouting of the foundation shale layer. In addition, during construction, it was necessary to provide protection of the shale, which tended to expand and deteriorate when exposed to the air. The Pittsburgh Testing Laboratories evaluated the

Figure 5.8. This shows the Wheeling and Lake Erie Railway relocated. (Project Maps, US Army Corps of Engineers, 1947)

compression and shear strength of selected core samples of the limestone, the black carbonaceous shale, the grey carbonaceous shale, as well as the sandy silt shales. Sliding friction tests completed the investigation.

In addition to the study of sliding and foundations bearing capacity, the research also considered the possibility of overturning induced by horizontal water pressure against the back of the dam. The study also included specifications for the concrete. This analysis of the Dover Dam provided the basis for detailed specifications and drawings used as contract documents and as a guideline for the contractor when building the dam.

No sooner had construction begun at Dover on January 6, 1935, than vociferous protests were lodged by the citizens of Zoarville, fearing their historic town would lie under the level of the Dover Reservoir. Zoar had been settled early in the nineteenth century by a religious sect of Zoarinites who had disappeared by the time the construction of the Dover Dam was underway.

Figure 5.9. A photograph of the Beach City reservoir and relocated Wheeling and Lake Erie Railway. (US Army Corps of Engineers)

By all accounts, it was one of the most successful communistic settlements in the New World. At first it appeared that the village of some forty houses would have to be moved to higher ground or razed. As protest mounted, not only by local citizens but by historical societies and other groups, a protective levee design evolved that would protect Zoarville from the design flood with a comfortable margin of safety. Another important factor in the solution involved what to do with nine producing gas wells that would be submerged under the waters of the proposed reservoir, creating a long-term potential environmental hazard.[14]

The so-called Zoar problem reached another level with an inquiry from U.S. Senator Robert J. Bulkley to Major General Edward M. Markham, chief of engineers, calling for his "careful consideration and advise as to what may be done."[15] Like a competent bureaucrat, Pillsbury delegated the problem to the Ohio River division engineer to reply in this matter.[16] Division Engineer Lieutenant Colonel F. B. Wilby, acting chief of engineers, indicated to the senator that the best solution was to build a levee to protect the village. In an effort to convince the senator, he sent a copy of the landscape architect's plan. He ended the letter by stating:

> I take pleasure in enclosing, herewith, for your information a photostat of the drawings prepared by the landscape architect showing how the village of Zoar will appear with the proposed improvements completed. I believe that you will agree with me that this photostatic study shows that the proposed works will not mar the appearance of this village.[17]

The levee was built, and today it still stands as a guardian of the historic village of Zoarville. In like manner, the small village of Tappan, population 500, was threatened with occasional inundation from Tappan Dam and reservoir. Unlike Zoarville, however, it was decided to relocate the entire village to higher ground.

Land Acquisition

With 54,000 acres in the project area, the Muskingum Watershed Conservancy District faced the daunting task of acquiring property below the pool level by purchase or condemnation. In other cases, flood easements had to be secured and the land cleared of all structures below the "plan flood" and above the normal pool level of the various reservoirs. A number of utilities, including roads, railways, gas lines, and electric transmission lines, were of necessity relocated with appropriate rights-of-way secured by easements and, in a limited number of cases, outright purchase.

In order to expedite the engineering work, as soon as the contract was signed in 1934 between the U.S. Army Corps of Engineers and the Muskingum Watershed Conservancy District, the district sought release of $2 million pledged by the state of Ohio for land acquisition. The first of four installments was received in July of that year. A priority system was established, placing securing land at the dam sites first, followed by those areas most likely to be affected by the operation of the dams.

The work advanced quickly with 500 parcels of land acquired by March 1935 at a cost of $1,500,000. By the end of the following year, 1936, nearly all of the first-priority land had been acquired, including thirty-four properties acquired by eminent domain. A further eight were awarded as the result of the verdict of juries.

This intense activity of the MWCD resulted in the expenditure of $4,200,000 for land purchases, contracts for bids, rights-of-way, and easements. An estimated $2,808,000 would be required to acquire additional land. It was reported that:

> The objections and protests arising from the hearings on the Official Plan were mild compared to those arising from the appraisal of damages and benefits. The District's board of appraisers had to deal with some 32,000 individual properties, including farms, city residents, business establishments; almost 300 industrial properties; nearly 300 miles of railroad tracks; holdings of 33 public utility companies; and many mining and quarry properties.[18]

Stored in New Philadelphia, Ohio, literally thousands of records of the MWCD testify to the magnitude of the bureaucratic and legal task undertaken. Relieved of the responsibility of dealing with the public in terms of land acquisitions, the army engineers could focus on the construction aspects of the project. In retrospect, the arrangement worked surprisingly well.

Clearly it is not necessary or even advisable to list all of the transactions undertaken but rather to present several case studies concerning individual land parcels to indicate the kinds of transactions undertaken by the Muskingum Watershed Conservancy District. The following three case studies were all taken from the files on the Tappan Reservoir area and are typical of the hundreds of similar cases throughout the watershed.

E. T. Haines Case

Purchased for $10,300, the farm of E. T. Haines and his wife Lona P. comprised 110.12 acres. The contract was executed on June 3, 1938, with a provision that the owners were permitted to remain on the property until April 1, 1939, rent free, but were restrained from planting any crops that would mature after April 15, 1939.[19]

As the owners of the farm of the former Haines property the district, for the sum of $6,864.00, granted an easement on August 1, 1941, to the U.S. Army Engineers for the right to flood the land in connection with the operation of the Tappan Dam.[20]

Writing to William Hines & Son, insurance agents, on April 4, 1940, H. P. Curtis of the district advised him that the property of the former E. T. Haines farm sustained severe wind damage to the barn, which was declared a total loss. In response Vance Hines of the insurance company supplied a check in the amount of $902, which represented the district claim.[21]

In a letter from H. P. Curtis of the land department of the district dated June 27, 1941, Roy James (a temporary renter) was advised that it was necessary to vacate the Haines house since it was lying in the right-of-way of the proposed highway relocation. It gave him thirty days to relocate.[22]

Earlier, on February 17, 1941, James was advised that although they had received a commitment for him to pay his delinquent rent on the Tappan area residence, the district had not received payment since August 8, 1940. As a result, if money was not received by March 1, 1941, the lease would be canceled and "the necessary steps taken to collect the amount due."[23]

During the same period, a number of local men bid on the Haines farm buildings with the stipulation that each successful bidder had to remove the structure in question by June 1, 1940. These contracts were for a small fraction of the appraised value and included the dwelling, barn, summer house, poultry house, garage, stable, and seven other miscellaneous buildings.

Mr. L. W. Edgar of Cadiz, Ohio, bid $50.00 on the barn, but as of October 8, 1940, he had not removed the structure as stipulated. Thus, H. P. Curtis of the land department of the district advised him that if his work was not completed by October 25, 1940, the district would take the purchase price and deposit and make other dispositions of the material. In a similar vein, Mr. Albert Willis received a letter from H. P. Curtis on October 8, 1940, that he had to clean up the barn on the south side of the road at the Tom Haines place, or the district would be forced to retain the deposit and do the work for him. All work must be completed by October 25, 1940. He had earlier bid $10.00 for the lumber from the barn on the Haines property and $3.00 additional as a deposit for cleaning up all lumber and trash on the same site. The $3.00 was to be returned when the cleanup was completed. These minor contracts were undertaken to clear the land in preparation for the operation of the Tappan Reservoir. The extension of these minor contracts for five months seems to be quite generous on the part of the district.

In a number of cases, roads and railways had to be relocated to bring them to an elevation above the "plan flood" level. Regarding the former E. T. Haines property, which had been purchased by the district, an easement was granted by the conservancy on October 21, 1947, in consideration of $1.00 to the Board of Commissioners of Harrison County, Ohio. This easement allowed the commissioners to relocate U.S. Route 250 to higher ground.[24]

J. Norris Stroud Case

The tale of acquiring the J. Norris Stroud property begins with a contract between J. Norris Stroud (and Hilda Stroud, his wife, and a second party, William H. Smith and his wife Emma) and the MWCD signed on July 1, 1935, in consideration of $19,000. The property in question consisted of two parcels of land. The first and larger contained 186.7 acres, while the second (much smaller) tract consisted of 20.13 acres. The district agreed to pay in three installments: the first $1,000 upon execution of the contract, the second thirty days later, and the remainder on or before September 1, 1936. If insufficient funds were secured through bond issues, no additional interest would be charged due to a delay of the final payment. In any case, the full purchase price would be paid prior to the operation of the Tappan

Dam. Before the first payment, the government had the right of access for construction work, while the owners could continue to occupy and use the land for agriculture and pay any taxes owed during the period of occupancy. The Corps of Engineers could undertake such activities on the property as surveying, relocation of pipelines, telephone and telegraph lines, and making minor road changes.[25]

As in most of the land-acquisition deals, the district was concerned about ensuring a clear title and in this case hired lawyer Barclay Moore to undertake a deed search. Starting with Sarah and Abraham Layport in 1839, the title had passed to William E. Clamblett, thence to James Fisher, who conveyed it to Elizabeth Whittaker on December 22, 1881. Elizabeth died in 1885 or 1886. James Fisher was granted the use of the property for ten years. James was a brother of Elizabeth Whittaker. The land thence passed to her heirs and thence to the Strouds' and the Smiths' families. In a curious move, a second contract for only $1,000, payable to the owners, was executed on the same day as the first, July 1, 1935. The property, known as the first and second tracts, described the same parcels of land as in the previous contract. The document confirmed the deed passing to the above parties on May 10, 1930. In addition, the contract stated that the title conveyed to the district was clear, free, and unencumbered with the exception of easements of the East Ohio Gas Company and an oil and gas lease dated June 17, 1935. This supplemental contract was apparently necessary in order to satisfy the district that it had a clear and free title to the property. This is indicative of the title searches that had to be done on most parcels of land.

In an effort to clear the property of a house, barn, and other outbuildings lying below the planned reservoir level, the district accepted a bid of $700.00 from Stewart Henderson with the provision that all of the buildings be removed by July 1, 1944. It was noted at the time that the contract price represented just twelve percent of the appraised value.

Finally, an easement for transmission lines was issued between the district and the Ohio Public Service Company, dated November 10, 1943. The easement included not only the Stroud et al. property, but eight other parcels of land. The easement through these properties was 150 feet wide. The company securing the easement would have to accept any claims for damage by outside parties.

Even without the complication of extended negotiations and litigation, this case indicates the effort required on the part of the district, land department, and lawyers in acquiring land for the flood-control project.

George Henry Case

Not all of the cases involving the public revolved around land ownership but nevertheless took valuable staff time and a court case to resolve. A seemingly trivial case resulted from George Henry building an access road across district property to connect his adjacent land with State Route No. 25. While there may not have been another similar case, there were many similar incidences to keep the land office busy. A number of these dragged on well past the completion of the construction and raising the pool levels in various reservoirs. In fact, there are several noteworthy cases that are still unresolved. The Henry dispute was not settled until November 1955. The earliest records begin in September 1949 with the deposition by William D. Moore to the effect that George Henry approached him to use his bulldozer to smooth out a track leading to his cabins from State Route No. 25. Moore undertook the earthwork on August 14, 1949. The excavation consisted of smoothing out a 900-foot-long roadway as well as some grading around his cabins.

On the same day as the Moore deposition, September 14, 1949, Ralph O. Wright of the district wrote a memorandum to Secretary Bryce Browning revealing that the original access road was built about the year 1880 between the properties of Dr. John Morrison, owner of the MWCD property, and John Spray, former owner of the George Henry land. This road was on higher ground and was not part of the dispute. Wright stated the crux of the matter:

> This access road as so located and used is still used by Mr. Henry and has not been affected by any changes in roads due to the construction of the Tappan Dam and Reservoir.

> The use of the District's lands to which the District objects is the establishment by Mr. Henry of a new access road which he intends to use, in addition to the original access road, for the purpose of developing certain cabin sites on the Henry lands. The District takes the position that the reservation or exception in the deed dated March 6, 1912, recorded in Volume 71, page 217,

refers to an established way which has not been affected by any operations of the District, and that the establishment of an additional way over lands of the District is unauthorized.[26]

The next move by the district resulted in the road being barricaded at the entrance of the disputed roadway with the state highway. To ensure that this access would not be used, the district had several transverse ditches dug across the alignment of the road.

Considering the move on the part of the conservancy as a hostile act, George Henry took the matter to the Court of Common Pleas of Harrison County, Ohio, on March 27, 1951.

The essence of the plaintiff's case was the claim that for more than twenty-one years, he and his predecessors had used the roadway as presented in 1951. The claim was based upon the following:

> That said roadway is used for the purpose of ingress and egress to Plaintiff's above described property and that the use thereof for a twenty-one (21) year period or longer immediately past has been open, visible, continuous, uninterrupted, and adverse, and that as a result thereof this Plaintiff claims a prescriptive right to the use of said roadway for the purpose hereinabove set out and especially as a means of access to Harrison County Highway #25.[27]

Further significant information surfaced in an earlier inter-office memorandum of the district dated December 4, 1941. The roadway in question followed the track of an access used by the U.S. Army Corps of Engineers, in August 1941, to provide access to farm buildings on the property then owned by the district. These buildings were later removed according to the stipulations that all man-made structures below the maximum "pool level" were to be removed. The track up to the farm buildings formed the first part of the new access roadway to the Henry cabins. William Moore, with his bulldozer, improved and extended the alignment of the track already prepared by the Corps of Engineers.

Additional evidence corroborated the claim of the Corps of Engineers' role to construct an access road in 1941. The evidence needed was a 1938 aerial photograph from the Soil Conservation Service that clearly revealed that no road existed, thus disputing the plaintiff's claim of twenty-one years use by him and his predecessors.

In a surprising move Judge Moore ruled against the district and in favor of George Henry. An interoffice memorandum dated May 24, 1955, confirmed the judgment. An appeal resulted in a judgment recorded in 1955:

> It is, therefore, the order of this Court that the previous judgment entry, filed the 24th day of May, 1955, and recorded in Volume 6, Page 464, of the Journal of this Court, be set aside and vacated, and further, that this order be now entered by the Clerk upon the Journal of this Court as of the date of the 24th day of May, 1955, and that this order stand as the final order of this Court from that day hence.[28]

It appeared that this was the end of the matter in favor of the district, but a letter of November 2, 1955, revealed that Henry was still using the access road, despite the fact he was forbidden from this use, and that he also had access to the property by another entrance.

Although not a matter of land acquisition or securing easements, the Henry case illustrates the time and effort necessary on the part of the staff of the district to secure its unquestioned right to use the land for the conservation project. With fourteen reservoirs, these case studies typify literally hundreds that had to be resolved for the district to satisfy its responsibility for land and easement acquisition.

The district's comprehensive approach to flood control, conservation, recreation, and agriculture was enhanced by visionaries seeking more than afforestation and saving historic villages. In an article dated February 3, 1935, George Hebard Maxwell, executive and educational director of the National Memorial Institute in Washington, D.C., reported on the spread of a homecroft movement, which envisaged 100,000 industrial workers living on homesteads and practicing a kind of subsistence agriculture while still working in the industrial sector. This would be made possible by the grand plan. This "back-to-the-land" movement could be enhanced by the comprehensive flood-control and conservation program underway in the Muskingum Valley.[29]

Major Arthur announced that bids for the Senecaville Dam would be opened on April 2, 1935, but cautioned that no contract would be awarded unless the MWCD was able to purchase the necessary land. The problem involved the price of land, which the conservancy deemed too high. Following a conference on land values, Congressman Robert T. Secrest appeared confident that Seneca Lake would be built as one of the fourteen dams in the watershed, but the prices for some of the properties was beyond the limit established by the MWCD. Nevertheless, he felt property owners would reach an agreement. In the same newspaper article, the conservancy decided to eliminate the Freeport Dam and Reservoir from the list of dams to be built because of exorbitant land values.[30]

An important breakthrough occurred on St. Patrick's Day 1935, when the conservancy secured 300 acres from three key farms. It appeared that the opposition deflated, and the remaining 2,500 acres would be acquired in a timely fashion. Earlier, Arthur had declared Seneca Lake would be the largest and most beautiful lake in the system. Construction on the dam began on May 10, 1935; two years later, on May 14, 1937, it was completed and ready to impound water.[31] Copies of the specifications for the Senecaville Dam and Reservoir were accompanied by drawings and site plans as part of the contract documents, in compliance with both the Corps of Engineers' and the Public Works' administrative regulations. From the table of contents, one can glean information on the various topics included in such specifications. This set of specifications is typical for all of the rolled-earth dams in the watershed.

In a similar manner, the specifications for the Dover Dam, a concrete gravity structure, were received for review in the Corps of Engineers' office in Washington on April 8, 1935, forming a companion to the Senecaville specifications. This was the only concrete dam in the project. The items covered in the specifications are informative, especially those concerned with concrete production.[32]

At the end of the month, April 28, Lieutenant Weinert of the Zanesville engineers' office announced that 437 men were at work on those dams that were under contract. George M. Brewster was awarded the contract for the Mohawk Dam and began to mobilize his forces and the erection of a field office for the Corps of Engineers. The Zanesville district engineer had issued an order to proceed with construction beginning on May 1, 1935. However, at the end of April, only fourteen men were engaged to work on the site.

J. A. Mercer, contractor for the Senecaville Dam and Reservoir, engaged his men and mobilized material on the site with the expectation of beginning construction on May 10, 1935. By working four shifts, E. J. Kiff Company excavated the area for the outlet work during the third week in April and placed 12,000 cubic yards of roll-fill material in the embankment. At the same time, the Ohio Highway Department completed the relocation of the road along the line of the dam. At this time, Kiff and Company employed 105 men, with work commencing on February 23, 1935, at the Wills Creek site.

The successful bidder on the Clendening Dam, N. Giavina, began work on March 4, 1935. Excavations for the east key wall and the toe trench were underway. In the "borrow" area the material was stripped for the embankment. In addition, excavation of the stilling basin and the downstream face of the dam was continued with approximately 5,000 cubic feet of rock fill placed on the downstream face of the dam. This work included thirty-three men.

Work began on the Tappan Dam on January 4, 1935. The contractor, Sammons-Robertson, continued removing earth from the "borrow pit," placing 8,000 cubic yards of impervious roll-fill material in the core of the dam during the third week in April. Work was begun at the same time on the toe trench on the downstream face, with approximately 4,000 cubic yards of earth moved during the week previous to April 28, 1935. Additional earth moving involved placing sand and gravel in the fill plus 5,000 cubic yards of rock in the toe trench. Rock excavated from the spillway area was also placed in the toe trench at the bottom of the downstream slope. The discharge tunnel was being trimmed and ready for concrete. In all, 102 men were set to work.

At the Beach City Dam, William Eisenberg & Sons placed approximately 12,000 cubic yards of rock fill as part of the blanket on the embankment, while excavation work at the outlet continued. Earth moving started on the relocation of Sugar Creek when the toe trench excavation had been completed. The contractor, under the terms of his contract, completed the U.S. Engineers' field office at this time. At the time of Weinert's report, the workforce at this site consisted of ninety-seven men. Work began on March 25, 1935.

As one of the earliest dams begun in the watershed project, concrete work was underway at the Charles Mill site by the end of April, where it was reported that 183 cubic yards were cast in the east retaining wall during the week preceding April 28. At the end of April, some 500 cubic yards of rock

were placed in the toe trench, together with 9,000 cubic yards of roll fill compacted into dike no. 2. Utilities were installed at the Corps of Engineers' field
office as required in the contractor's contract. For the week preceding April
28, sixty-six men were employed.[33]

In a vote of eleven to three on June 8, 1935, the judges of the Court
of Common Pleas, which comprised the Conservancy Court, approved the
construction of the Bolivar Dam and Reservoir. In the meantime, the army
engineers had prepared contract drawings and specifications and were scheduled to award the contract on June 19. At the same session, the court eliminated the Freeport Dam from further consideration as one of the MWCD
dams.[34] A newspaper article featuring Weinert's weekly progress report on
June 9, 1935, noted with pleasure that ten dams were under construction
and a thousand men employed. Since the primary objective of the New Deal's
Public Works Administration was to provide jobs for the unemployed, this
news was undoubtedly gratifying to both state and federal agencies. Weekly
reports were published in a similar form at each site, giving details of the
progress. Weinert announced that Bates and Roberts, successful contractors
for the Dover Dam, received notice to proceed with operations at that site.[35]

Between May 2 and May 7, heavy rains covered the Muskingum watershed, resulting in an estimated loss of more than 2,500 tons of topsoil, but
because various dams were being built "in the dry," so to speak, the river
channel conducted water through the sites with little damage to the works.[36]

Again, from August 7 to 15, a summer freshet eroded six inches of topsoil
from the equivalent of 2,600 farm acres. The material was swept down the
Muskingum River into the Ohio. The Soil Conservation Service estimated
that there was a loss of one million tons of irreplaceable soil. Some damage
was sustained on the construction sites, but the work continued at a steady
pace.[37]

With conservation being an essential facet in the construction of the
watershed project, it was expected that such soil losses would be greatly
reduced with the flood-control dams, afforestation, and improved agricultural procedures such as contour plowing.

Captain A. C. Liebert Jr., in charge of operations for the U.S. Engineers,
reported on September 1, 1935, that good progress had been made during

the past week with 2,391 men employed, establishing a new record. Liebert reported on each of the dams under construction, as well as indicating that plans were underway for dredging part of the Muskingum River for navigation purposes, including removal of the Third Street Bridge piers in Zanesville, which were deemed to be a hazard to navigation.[38]

Earlier in the summer, the MWCD made application for enhancing the official plan in the amount of $13,912,000—the total for all of the projects. This included funds for settlement and rehousing of land owners affected by the construction of dams and reservoirs, five projects of soil-erosion control and reforestation, and the construction of eleven small-scale water-conservation reservoirs to supplement the larger dams and reservoirs. These presumably were to be designed by the Soil Conservation Service. Lastly, channel improvements and protection works at Massillon, Roseville, Newark, Zanesville, and Killbuck Creek were undertaken by the Corps of Engineers. The proposal, entitled "Application for Approval of Projects for Work Relief," was submitted to the federal government for approval on May 28, 1935. The report stated:

> The work requested will round out and complete a project already started which otherwise would be left incomplete and only partially effective. The work now underway will not reach its full value unless the other features in the comprehensive program are provided for. This work will help to solve problems of conservation of natural resources, land use, preservation of high standards of living, and recreation.[39]

On August 16, Mr. Gus Kasch of Akron, Ohio, sent a lengthy telegram to Secretary of the Interior Harold Ickes. The telegram's essence was to urge rejection of the Muskingum Conservancy application for an additional $13 million. Ickes, writing to Captain Lucius D. Clay of the Corps of Engineers on August 26, enclosed a copy of the Kasch telegram and a copy of his reply to Mr. Kasch. On September 4, Clay wrote directly to Kasch with regard to this matter. His final paragraph is worth quoting:

> The request submitted by the Muskingum Conservancy District has been analyzed by this office and as a result of this analysis it has been found that the work is not eligible for inclusion in the relief program as the cost per man year for the direct labor involved is considerably in excess for that established for the relief programs.[40]

Even the president became involved in this matter, and his special assistant, Thomas P. Carol, wrote to Kasch on August 26 indicating that his concern for the project would be given full consideration and the copy of his telegram sent to Lieutenant Colonel Glen E. Edgerton, civil engineer works application advisor for the War Department.[41] This and other public works during the New Deal caused ordinary citizens to write directly to the president of the United States stating their concerns.

The board of consulting engineers met and considered the movement of soil in the embankment at Tappan Dam. The group consisted of representatives of members of the board of engineers from the Division Office of the Corps of Engineers at Cincinnati down to the resident engineer at Tappan Dam. Consulting engineers in private practice and engineers from the MWCD were also present. The meeting was chaired by District Engineer Arthur, but the engineering information was presented by Mr. Knappen, chief of the engineering division at Zanesville. The first indication of movement occurred on August 26, 1935, but the next day, earth had moved 4.7 feet. The movement progressed rapidly to a maximum of 7.4 feet on September 1 at the upstream toe of the dam. After much discussion, including a site visit, the board reached a unanimous opinion that the landslide occurred at a level of six to nine inches below the contact of the rolled-earth embankment and the original foundation material. It was a classic shear failure developing a slip section at the bottom of the embankment. Before the meeting, a fill of "high-angle" material displaying high internal friction and with sufficient thickness to develop the necessary shearing resistance was installed to preclude any further movement. Started on August 31, the fill measured approximately ten feet in height and extended upstream forty feet from the toe. This expedient was thought to provide sufficient additional weight to prevent further movement of the embankment on the foundation. The board agreed on a new profile for the upstream slope of the dam, which would produce a satisfactory result. Neither the designers nor the contractors were implicated in this failure. The new profile has stood the test of time and has indeed proven to be "satisfactory." The new profile for both the upstream and downstream faces extended across the full width of the valley. Further, the board decided that the embankment should not be raised higher until the corrective measures were begun.

Concerned with landslides at other dams then under construction, the board reviewed the designs for the Clendening, Senecaville, Piedmont, Mohicanville, and Leesville embankment dams. The upshot of an extended discussion on the Clendening Dam was that the board recommended a wait-and-see approach, emphasizing the importance of daily checking at gauges and of elevations. It was clear, though with the one exception, that the board felt that no remedial action was needed at the time.

In viewing the design of the Senecaville Dam, the board requested that a rock toe be added to the upstream north embankment to preclude a failure similar to the landslide at Tappan Dam. In a further discussion the board included the south embankment in its recommendation for the extra rock fill to be placed on both up- and downstream toes of the dam.

With regard to the Piedmont Dam, Mr. Gerig of the board questioned the adequacy on the west side of the valley. The board unanimously agreed that the west side was adequate against a shear failure but recommended that additional rock fill be placed on the steeper slopes. However, no definite plan was approved.

In the case of the Mohicanville Dam, the board felt the design was adequate, but requested that the embankment be carefully observed for movement. Mr. Knappen assured the board that all of the dams would be closely monitored.

Although expressing concern over the adequacy of the Leesville Dam, the board reserved judgment until after inspection of the site.

The following day, September 15, the board visited the Mohawk site to study the foundation in the embankment. A cofferdam was constructed for the erection of form works in the conduit under the dam, since a "quick" condition had been observed. Colloquially called "quicksand," this condition arises when uplift fluid pressures act on a porous material, expanding the grains and causing a notable lack of bearing capacity. The foundation material was found to contain four percent silt, which, although a small percentage, was enough to cause its very low permeability. The cofferdam was under a six-foot head of water. In addition, the seventeen-foot-high cofferdam was bulging at one part of the toe, believed to be caused by "piping" (that is, open channels in the embankment). Board member William McAlpine recommended that a trench be excavated between the cofferdams to act as a drain and reduce the water pressure causing the quick condition. The board devised a plan for embankment construction. It was agreed that the foundation should first be stripped of all boulders and loose gravel, followed by a layer of approximately three inches of impervious material spread over the foundation and covered with the same amount of gravel from the outlet excavation. The gravel should be then rolled into the embankment's impervious layer. It was the opinion of the board that this would stabilize the foundation. They also recommended, in their modification, that the upstream slope should be flattened to preclude any type of landslide and also the stream channel modified below the dam.

Following this discussion, the board visited Dover Dam, had the opportunity of inspecting the foundation, and found it to be satisfactory after being pressure-grouted in the cutoff wall area.[42]

Deeply concerned about possible shear failures in several of the dams following the Tappan landslide, the board of consultants met again on September 19–20 to consider the situation further. The group assembled at Piedmont Dam to inspect the embankment and learn of the plan to add additional rock on the toes. After extended discussion, the group decided that the fill material on the downstream side should be entirely rock, whereas on the upstream face it should be a combination of rock and earthen material.

Next, the party visited the Clendening Dam and determined that additional fill should be placed at the toes of the embankment.

While visiting the Atwood Dam, the board determined that overburden would have to be stripped to rock under the main section of the dam. The final decision, however, would have to be made when the excavation opened up this geological area. As the group revisited Dover Dam, it was decided that all of concrete monolith no. 5 should be cast on the limestone, and that cracks in the limestone foundation should be grouted.

The following day, September 20, the board assembled at Mohicanville Dam and was joined by Mr. Belknap, engineer of the Loudonville area. The design reflected in the contract for the embankment was approved. As a precautionary measure, all excess waste material not needed was ordered to be placed on the upstream blanket to reinforce the toe of the embankment to ensure against a Tappan Dam-type failure.

The Charles Mill Dam posed a special problem: under the impervious section of the embankment, numerous springs were encountered. It was, therefore, decided to excavate under the pervious section down to bedrock and use pervious fill down to that level. At the earlier meeting of the board, McAlpine had recommended that a trench be cut down to the water table. The contractor had completed the work, revealing what appeared to be a stable foundation. Nevertheless, there was concern that piping might occur behind the gravelly material and the impervious layer. As a precautionary measure, it was decided to drive steel sheet piling across the bottom of the dam. In addition, a 42-inch-diameter inspection hole would provide a means of monitoring the situation.

Wills Creek was the last of these site visits by the board. The group found that a section of the embankment near the terrace top was to be built up of earth containing a very large portion of fire clay. At that location it did not pose a problem; however, it was considered a dangerous material to be used in an embankment, therefore it was used only in very small quantities and mixed with other materials.[43]

These visits of the board of consultants illustrated clearly the important role of experts in addressing embankment problems encountered in actual construction of the various dams.

Because of structural problems at Tappan Dam and potential local shear failures at a number of other sites, remedial measures were undertaken, and work progressed without delay. By early October operations at Mohawk Dam and Senecaville Dam were underway. At the same time, the construction schedule at Wills Creek necessitated the employment of 247 men, while an additional 121 men were at work on the Pennsylvania Railroad relocation. Captain A. C. Liebert presented details of work at each of the fourteen sites in his weekly report. With this level of activity, it must have been an awesome site to see more than 2,000 men and earth-moving equipment in action. At the same time, the Corps of Engineers was supervising repairs at Lock and Dam No. 11 on the Muskingum River above Zanesville. This work included repair and reinstallation of the Boulé movable dam.[44]

At the beginning of January 1936, District Engineer Arthur summarized the progress made in the calendar year 1935. These estimates were presumably based on quantities of material, both earth and concrete, based on the following completion percentages:

Name of Dam	Percent Completed	Name of Dam	Percent Completed
Tappan	91	Senecaville	35
Charles Mill	89.8	Piedmont	42.7
Wills Creek	40.8	Mohicanville	26.2
Clendening	59.4	Pleasant Hill	3.1
Beach City	29.4	Leesville	21.7
Mohawk	27.5	Atwood	9.8
PA RR Relocation	22.6	East Ohio Gas Line Relocation	71.7
Dover	19.5	Bolivar	
PA RR Relocation	74.2	East Ohio Gas Line Relocation	8
Zoar Levee	27.6		9.9
W&LE Relocation	10.6		
B&O Relocation	22.8		
Gas Lines	9.9		

At the peak, 4,300 men were employed per day.[45]

Early in the new year, January 23–24, 1936, the board of consulting engineers met once again to review the work underway at the various dams and reservoirs and to address any problems encountered in the construction work. Amongst those present were Gerig and McAlpine of the chief of engineers' office; engineers Justin and Hill; private consulting engineers, together with representatives from the conservancy district; and engineer Knappen and four of his colleagues from the engineering division of the Zanesville district office. The group also included the area engineer and, since Atwood Dam was on the agenda, the resident engineer at that site.

Mr. Knappen presented two plans for a change in the location of the Atwood spillway. The board strongly endorsed the change and selected the second alternative, which consisted of placing the spillway channel completely in a cut. They eliminated the projected hydraulic jump at the inlet and in an effort to prevent scour endorsed a 200-foot, concrete-lined channel stretching well beyond the gas lines crossing the site.

In constructing the abutment, Mr. Knappen said, "the question at hand was whether the excavation in the valley floor should be extended down to the rock layer before placing the rolled-earth embankment." Since the material had a twenty-five-percent clay content, which provided a nearly impervious material, it was recommended that the material be left in place.

The results of a model study carried out in the soil laboratory to represent the shale that would be encountered in excavating for the spillway cut were presented. Apparently the model had been under test for several months and at that stage had shown no appreciable leakage. At that time, another revision was placed on the table for the consideration of the board. This involved the main embankment, where reducing the rock toe and placing a "dump fill" riprap on the upstream slope was contemplated because it was difficult to find suitable rock for riprap and toe rock at this particular site. Concern was expressed with regard to possible leakage and even piping in the interface between the pervious and impervious layers. The board, therefore, reached the opinion that materials should be graded from fines in the impervious section to larger rocks in the downstream face of the rock toe. At this point in the meeting, the board adjourned to visit the soil-mechanics laboratory to view the model tests.

In an amusing discussion, McAlpine questioned why the Beach City spillway design should be reduced in capacity. He was reminded that it was necessary to incorporate this reduction to meet the requirements previously approved by the board. No further reference to the Beach City Dam appears in these minutes.

Model studies played a significant role in seepage studies for the Bolivar Dam. The model study revealed seepage would quite likely occur in the terrace area even if an earth blanket were placed on the upstream face. It was believed it would reduce the seepage by only about sixteen percent. In the case of the Bolivar Dam, the model studies of the embankment section of the narrow portion revealed that seepage could be expected on the back slope of the terrace. At this point, the board felt there was a need for a rock toe on the back slope to prevent any sort of landslide occurring caused by seepage through the terrace. In addition, the board recommended that a rock fill be placed on the terrace area extending some 400 feet from the toe. In regard to the Tappan Dam, which in many ways started the board in its broad investigation of possible landslides on all of the dams, Mr. Knappen reported that the field observations bore out the theory he had presented earlier to the board. No more movement in the embankment could be detected. The board spent a considerable amount of time discussing Mr. Knappen's modification of the Terzaghi theory. No action was taken endorsing this modification of a well-recognized procedure for curtailing seepage in the foundation and embankment areas.

In the case of the Leesville Dam, changing a planned grout curtain under the center line of the dam was proposed. This curtain would be tied under the key wall and thoroughly grouted around the hill, the tunnel, and the intake tower. This would be an alternative to grouting downstream from the center line of the dam. In addition to approving this procedure, the board suggested that three grout holes be placed in the invert of the tunnel at the center line of the dam and thoroughly grouted. In constructing a tunnel, one has the choice of boring through, like a mole, or making an open cut and later covering it over. It was this latter method that reached the board as a proposal from the contractor at the Mohawk Dam. They could hardly object because the proposal met the conduit details shown on the plans. Again, to improve the hydraulic behavior of the embankment at the Pleasant Hill Dam, the board considered a proposal for making a transition between the pervious and impervious layers in the dam. The board approved a change, suggesting that the sandy rock be thoroughly watered to wash down the sand into the rock, starting with silty sand material, which is nearly impervious, at the impervious section, and graded impervious materials against the interface between the transition and the pervious layer.

Having been informed that the foundation at this dam was satisfactory and consisted of well-graded materials, the board took no further action on the construction at Pleasant Hill Dam.

With an extensive agenda, it was necessary for the board to meet a second day, on January 24. The first site considered was the Senecaville Dam. The engineers knew during the design phase that there was a four-foot layer of indurated clay, that is, clay that had been heavily compacted but had a proclivity for expanding slowly under moisture conditions. Eliminating this potentially damaging material involved lowering the foundation's elevations. Further change was considered: a modification of the cross section by placing a rock filter on the surface of the foundation extending from the upstream toe to the impervious section of the upstream rock toe. This filter would be covered with impervious material available on-site and protected by a rock paving. Mr. Gerig, who frequently questioned proposals made to the board, expressed his concern that the filter would form a piping situation in the dam, but he yielded to the overall opinion of the board.

While there was no serious problem at Clendening's Dam, the gauges showed that the actual settlement was 0.42 feet and that it was practically negligible at the toe, being 0.22 there. Board member Mr. Hill was concerned that there was too little settlement rather than too much, but it was pointed out to him that the rock fill was placed carefully, and this condition was satisfactory.

In a similar situation, the board received information that the Piedmont Dam had experienced practically no movement, and the embankment was being built in conformance with the wishes of the board. Though no specific data were presented, the engineers reported that readings on the settlement gauge indicated that the foundation was consolidating in line with the theoretical values computed for the structure. Arthur was of the opinion that it was possible to predict settlement in the foundations, and the results bear out the design computation of his engineering division. Although little information was presented in the official minutes, it appears that a photoelastic study, in addition to a study in the soils lab, was undertaken to determine the stresses distributed in the foundation on Tappan, Clendening, and Piedmont Dams, and that they checked within ten percent of the field values.

Turning their attention to the Wills Creek Dam, the board was presented with details of revised slopes of the embankment section at both the terrace and the stream bottom. In the revised plan, the engineering division's design followed very closely the recommendations earlier made by Mr. Gerig. The board approved this change. In order to allay any concerns with regard to shear failures, Mr. Knappen pointed out that tests on the foundation itself showed that the stress level was so small that it was felt impossible to have a landslide as witnessed at Tappan Dam.

As a last agenda item, the board reconsidered information with regard to Atwood Dam. It felt that the material in place could remain and did not have to be removed in the valley bottom. They also recommended that the concrete key wall be omitted but pointed out that it would not result in material savings in the contract. With this discussion completed, the board adjourned.[46]

Just before the Board of Consultants met in January, the local newspaper proudly announced that $3.5 million had been allotted to the Muskingum Watershed Conservancy District.[47] The credit was given to Congressman Robert T. Secrest. The engineers' office at Zanesville justified the necessity for additional funding on delays and damages caused by the August floods that proved expensive; equally important, unforeseen conditions in foundations at several of the sites and the necessity of increasing dam sections to provide greater safety required additional funds. It is clear that the modifications approved by the board resulted in additional costs. The emphasis on safety resulted from the landslide at Tappan Dam. The article further said that if the additional funds had not been received, the plan would have been curtailed, or an additional $1.5 million tax would have to be imposed by the Conservancy District. The speed with which the money was received is quite noteworthy. It was only the week before, on Thursday, that a delegation consisting of Brice C. Browning, secretary of the Conservancy District; the chief legal counsel Robert N. Wilken; Charles Spencer of Newark, Ohio; and Major J. D. Arthur Jr., district engineer from Zanesville, met in Washington. Within a week the funds had been secured.

Writing from the Office of Chief of Finance of the War Department, Major L. T. Worrall wrote to the chief of engineers, enclosing a copy of a letter of January 27, 1936, from the federal emergency administrator of public works to the secretary of war, stating that the $3.5 million in additional funding had been made to the War Department, Corps of Engineers (flood control). The funds were placed to the credit of the War Department under the 1933–1937 appropriation of the National Industrial Recovery Act.[48]

On February 10, 1936, Arthur submitted an estimate dealing with the navigation system on the Muskingum and a cost estimate for the Dillon Reservoir. It was apparently questioned by the Ohio River Division in Cincinnati, causing some delay, since the Rivers and Harbors Office did not receive it until September 21. The estimate for renewal of the navigation system on the Muskingum amounted to a staggering $33,998,000. The estimate for the Dillon Reservoir, which would form the fifteenth dam in the conservancy district, was estimated at $7,300,000.[49] This is the first reference to cost estimates for the Dillon Reservoir.

Secretary of the Interior Harold Ickes, administrator of the New Deal-era Public Works Administration in Washington, wrote to Major General Edward M. Markham, chief of engineers, with regard to the Muskingum Watershed Conservancy.[50] It was a request for an additional $4 million. In blunt language, the letter stated that if this money were not forthcoming, the district court would have to dissolve the conservancy district. The indication was that the district was at the limit of its resources because of cost overruns on highway relocation and high damage awards. Ickes proposed that a committee be appointed to go to the district and send a report to him. The report covered the reasons for the overrun in estimated costs and questioned whether the $4 million would be sufficient to complete the project. It questioned the progress of the project as a whole, including that part of the project being constructed by the district. The report sought information on the quantity of work that had been done, how much remained to be done, what agencies—local and federal—had participated in the project's cost, and what percentage each had contributed. With this report, sufficient information should be available to make a decision on additional federal funding. Administrator Ickes also suggested that a representative of the War Department, meaning the Corps of Engineers, be on the committee and be involved in the preparation of the report.[51]

The district engineer was to submit his report according to the requirements listed in House Document 308 of the 69th Congress, authorized in 1928, which required the Corps of Engineers to survey more than 180 rivers and a number of unnamed tributaries with a budget of $7.3 million.[52] At the time Congress passed this act, Major General Harry Taylor of the Corps of Engineers stated it would have a far-reaching influence in controlling and coordinating all works connected with the diverse beneficial uses that might be made of the streams under federal jurisdiction. Taylor urged that the work be undertaken as soon as possible. In compliance with the Flood Control Act, a 308 Report dated December 1, 1934, was prepared for the Muskingum River. It was revised on January 10, 1942. Colonel R. G. Powell, division engineer of the Corps of Engineers, wrote to the Ohio River Division on August 26, 1936, requesting information as to when the 308 Report on the Muskingum River, which was returned for revision in January 1934, would be ready for his review.[53]

Major R. G. Moses, writing for the president of the Mississippi River Commission, addressed Colonel Graves in a cover letter forwarding the report of the division engineer of the Ohio River division, prepared in accordance with House Document 308. In addition to the report, Major Moses reminded the reader that in 1934 the Mississippi River Commission had acted upon the 308 Report.[54] A copy of their action was included. Brigadier General H. B. Ferguson, president of the Mississippi River Commission, replied, making the following points in endorsing the report:

- The requirement for navigation appeared to be provided for under the current system.

- It was not suggested that water power could be developed economically.

- Completion of the fourteen flood-control dams together with other flood-control structures and improvement provided all of the flood control that could be justified economically.

- The storage capacity of the proposed Dillon Dam primarily for reducing flood stages in the Ohio and Mississippi Rivers at Cairo, Illinois, would result in the reduction of the flood elevations of 1/10 of a foot. It was believed that such a benefit would come at a fairly high price.

- There was never any suggestion by any of the agencies that irrigation was desired.

Earlier, Colonel Pyle had forwarded the copy of the revised 308 Report, dated September 10, 1936, to the chief of engineers.[55] The revised report of 1942 provides an excellent summary of the entire Muskingum Watershed Project and will be referred to in the conclusions.

Floods Punctuate the Progress of Flood-Control Development

Destructive floods have served to stimulate nationwide studies of engineering means to combat the power of floodwaters. These had a direct influence on the Muskingum Watershed project. To begin with, the disastrous 1913 flood served as the benchmark and was the basis of the official plan flood upon which all of the design work for the fourteen dams and reservoirs was based.

In 1927, the lower reaches of the Mississippi sustained a massive flood that resulted in the promulgation of the House of Representatives' Document 308 and the Flood Control Act of May 15, 1928. The 308 Reports provided, for the first time, a comprehensive view of navigation, flood control, hydro-electric power generation, and irrigation. These reports served as comprehensive planning and development documents for a multitude of rivers. This 1927 flood of the lower Mississippi ended the Corps of Engineers' stubborn championing of the levees-only policy. Even with a more enlightened view on the matter, the Corps of Engineers was still wedded to a mission focused on navigation only, and flood control and other aspects of the 308 Reports had to be couched in terms of benefits to navigation.

The Ohio flood of 1933 prompted Huntington, West Virginia, and other towns on the Ohio to petition Franklin Delano Roosevelt for flood relief, not only on the Great Kanawha and Muskingum but also to address the problem of flooding on the mighty Ohio itself. This initiative resulted in the founding of the Ohio Valley Conservation Congress. Ultimately it provided protection behind flood walls and levees for Huntington, West Virginia; Portsmouth, Ohio; and other communities. The overall problem of flooding, however, was not addressed.

The 1933 flood clearly influenced the 308 Report of 1934 prepared for the newly founded Ohio River division. This early version of a 308 Report recommended not only many local protection projects but an unprecedented thirty-nine reservoirs on various tributaries of the Ohio River. Since the Muskingum Watershed Conservancy was already founded, it was not included in this 308 Report.[56]

Beginning on March 9 and extending until March 22, 1936, two massive rainstorms enveloped the Northeast from the Ohio Valley into New England

and as far south as Washington, D.C. From ten to thirty inches fell on ground covered with snow, which added to the runoff. Pittsburgh, in particular, was very hard hit. The $43 million appropriation by Congress to ameliorate the damage really addressed only the symptoms and not the root cause of the problem. With flooding in Washington, and not isolated to New Orleans as in the 1927 flood, Congress was finally stimulated to do something about this chronic problem.

The challenge was taken up by Senator Royal S. Copeland. He was the champion who guided an omnibus flood control bill through Congress, which became the Flood Control Act of June 22, 1936.[57] This ended the charade of the Corps of Engineers justifying all flood-control projects in terms of the benefits to navigation.

Less than a year later, and before any remedial flood-control projects could be undertaken, the Ohio River suffered another destructive freshet that, in terms of damage, exceeded the 1936 flood by one order of magnitude.[58]

Now, in mid-summer of 1936, the Corps of Engineers, under authorization of the Flood Control Act, could become a full partner in the Muskingum Watershed Project and address flood-control measures on the Ohio River and lower reaches of the Mississippi as part of their mission. The fourteen dams were thus seen in a broader context than just part of the watershed area. In fact, this concern was the most powerful justification for the Dillon Dam, completed in 1961, which became part of the watershed system.

In a letter to the chief of engineers from Colonel Powell, division engineer, the colonel provided a succinct status report on the Muskingum Watershed Project. Writing on January 20, 1937, he noted that with all of the dams under construction and expected to be completed by the end of the year, the staff of nearly 400 in the office had been reduced to 150. The field forces were still needed, however, to bring the computations up to date to incorporate any design changes so that as-built drawings could be provided at the end of the project. Field forces were also needed for inspections. When this aspect of the fieldwork was completed, there would be a further reduction in personnel.[59]

At this stage of the construction, several items had not been advertised for bid, mainly the Magnolia Levee, the Sandyville Levee, the operator's quarters at Dover Dam, a railroad relocation at Beach City, and work at Massillon. With obvious delight, the colonel reported that Charles Mill, Tappan, and Mohicanville had been completed and accepted.

In addition, of the nine railroad relocations, six had been completed. At Dover, relocating the Wheeling and Lake Erie Railway was about 65 percent completed and expected to be finished by October 1, 1937, while the Beach City relocation of the W&LE appeared to be 90 percent complete, as contrasted to the B&O realignment that had just been started.

Relocation of public utilities kept pace with the other work and, at the time, was reported to be completed with one exception, which would be finished by July 1, 1937. Waiting for the W&LE realignment to be done before the Zoar levee would be finished resulted in a delay until the end of the year. Except for Dover Dam, Powell expected all other work to be completed by December 15, 1937. Hired labor, that is, labor engaged directly by the Corps of Engineers, was to be used to clear the land in reservoir areas. This work was contingent upon the MWCD securing land.

Meeting for two days on February 17 and 18, 1937, the board of consultants reviewed three projects still under construction, especially the reported movement of the Clendening Dam.[60] Setting the stage for discussion, Major Lieber presented details of the progressive lateral movement and settlement of the embankment. After inspecting the dam, the board recommended that test pits be located at four places in the embankment, one of the pits to be sunk through the rock fill at a point of greatest "bulge."

Concurring with Mr. Justin, Mr. Gerig endorsed his idea, urging that the embankment be secured against possible failure in the case of a flood. As a result, the district engineer ordered that the toe and upstream face between surveying stations no. 2 and no. 8 be loaded as soon as possible. Furthermore, additional fill would be placed across the entire face following the stabilization at the point of maximum movement.

After being briefed on a fault in the foundation at Dover Dam, the board agreed that the foundation under Monolift fourteen be carried down to a sandy silt shale layer, and when this layer was reached an investigation was to be undertaken to determine whether additional anchorage should be built.

Endorsing the proposed plan for grouting the east abutment of the Mohawk Dam, the board turned its attention to the more serious matter of the movement at Clendening Dam. Since considerable movement had taken place since the end of January, both Gerig and McAlpine recommended that the upper part of the rock face on the upstream slope be removed by hand as quickly as possible. Mechanical equipment was not to be used, while at the same time loading of the toe should proceed. If this remedial work did not prove to be successful, then the upper portion of the embankment was to be removed manually. Lieutenant Colonel J. D. Arthur Jr. approved this procedure. On the following day, Messrs. Gerig and McAlpine laid two further recommendations on the table. First, that the downstream face be lowered in addition to work on the upstream face to prevent any potential movement of the toe. The second recommendation, to be undertaken after the flood season, involved the removal of all earth above the impervious layer and its replacement with pervious material. The board approved these suggestions before adjourning.

Change Orders

Change orders occur in large public works with some frequency when the contractor of note claims the work in progress varies from the contract specifications and drawings, or unforeseen events beyond his control cause delays and added expenses. For example, the contract may call for excavation in soil but the contractor encounters rock, which is much more expensive to remove, thus causing a change order to be issued. Alterations in the contract by the client, in this case the Corps of Engineers, often initiated a change order by the contractor, claiming extra compensation and possibly an extension of the contract time.

To evaluate the change orders for all fourteen dams and other flood-control structures would yield little in the way of fresh incites into the history of the project; rather, it seems more enlightening to examine a selective set of change orders, principally at the Beach City Dam.

With the receipt of a letter dated April 26, 1937, concerning the construction of the Clendening Dam, contractors Boso and Ritchie undertook the remedial work required to satisfy the board of consultants' recommendation.[61]

Since the board felt that the contractor had placed undesirable material in the embankment, it would be up to him to put it right without additional cost, and that would be the end of the matter. Following a detailed investigation of settlement and bulging of the embankment, the Zanesville army engineer determined the root cause by discovering that the upper portion of the dam had been constructed using a shaley material, poorly consolidated, and already partly disintegrated and in a saturated state. As if that were not enough, severe cracks of the type associated with landslides were detected. This material was found to move both vertically and horizontally. Determining the material to be objectionable, Lieutenant Colonel Arthur quoted the specification, which stated clearly the contractor was to cut out and replace objectionable material and rebuild any section at his own expense.[62] In reply, the contractor claimed that there was no objectionable material in the meaning of the term used in the contract. This seems to have beeen a studied legal nuance, probably recommended by the contractor's legal council. Furthermore, no landslide had occurred "within the meaning of the specification." While Boso and Richie claimed that the district engineer had accepted the dam, the government disagreed. The order for the work to be undertaken to stabilize the bank was received on April 26, 1937, and constituted a change order, thus, according to Boso and Ritchie, the contractor should be compensated.[63] Arthur, writing to the division engineer, stated at the end of his endorsement:

> It is my interpretation of the wording "during the construction or after completion but prior to the final acceptance of the work" that even though the embankment was completed on 16 December 1936, final acceptance had not been made and, therefore, the contractor is obligated to perform the work contained in my order of 26 April 1937. As the work is specifically called for by the contract, there has been no change in the contract provisions and hence, no supplemental agreement nor extra work contract is justified. Paragraph 8–05 (H) covers the removal of objectionable material as the presence of objectionable material in the embankment and the slide resulted through no fault of the contractor, payment for the work will be made as provided in the specifications.[64]

At about the same time, a more favorable claim by the Gilbert Construction Company, which was involved in the relocation of the Wheeling and Lake Erie Railway track at the Beach City Reservoir, encountered a situation in which the foundation was composed of boulders and fractured rock and earth unsuitable for embankment use. The design, on the other hand, indicated that the foundation material consisted of suitable rock. A change order was approved for $3,006.27.[65] In another case, at the Beach City Dam, contractor William Eisenberg & Sons, Inc., sought a time extension because of delays by another contractor to complete the B&O relocation on time. Eisenberg sought both an extension of time and also compensation of $58,070.27. The Corps of Engineers was agreeable to the time extension but refused to provide extra compensation.[66]

While Beach City Reservoir contractor Eisenberg was granted a time extension, one would think that the Gilbert Construction Company (in charge of the relocation) would be fined for a delay in completing the contract. In claiming a time extension, Gilbert stated "the placing of the relocated line in service will be indefinitely delayed through failure of the W&LE Railway Company to perform its agreement with the government." The claim was neither for an increase in funding nor an extension of the contract schedule, but rather for a reduction of the amount of funds retained by the Corps of Engineers until final approval of the entire project. This is a standard procedure used in such cases. The amount withheld was specified in this case: "the total amount retained will never be less than 5 percent." With the final payment held indefinitely, the Gilbert Company requested that the percentage be reduced to 1.5 percent. Approval of the change order would mean that Gilbert and Company would receive a sizable payment for work done even before the completion of the contract.[67] The change order was approved.

By August the contractor, Gilbert, submitted another change order, this one requesting a time extension in the contract to enable the removal of 5,700 feet of abandoned railway track in connection with the W&LE relocation. Since the work was not covered in the original contract, the Corps of Engineers granted a seven-day extension with no increase in the contract amount.[68]

As indicated above, William Eisenberg & Sons, Inc., began a contentious issue with Lieutenant Colonel Arthur, the district engineer. Writing to the chief of engineers, the colonel claimed no knowledge of any claim by Eisenberg & Sons in connection with the Beach City Dam. Believing the contract to be complete, he stated that the final payment to Eisenberg was made on January 10, 1938.[69] Not content with the ruling of the district engineer regarding the use of unsuitable material in the Beach City Dam, the contractor claimed compensation or certain expenses and losses in connection with the ruling of the district engineer.[70]

Still not satisfied, William Eisenberg tenaciously pursued their claim for $58,070.79, protesting what they considered "grossly erroneous and arbitrary statements in the foregoing endorsement" by the district engineer. The letter from Eisenberg & Sons to Arthur gives details of the claim in four pages of carefully considered information on the situation. Unable to resolve the matter, the case ended up in the Court of Claims.[71]

Having decided to take the claim to the Court of Claims, the court stated on December 27, 1941: "You are hereby requested to furnish to the Court of Claims, that the same may be used as evidence on the trial of the above-entitled cause and pending in said court, any information or papers . . ."[72] Thus, after much delay, the trial was imminent.

In a flurry of activity, the War Department issued travel orders to Mr. E. R. Baughman on December 31 to permit him to assist in the preparation of the government's defense in the Eisenberg case, which was scheduled for January 20–22, 1942.[73] The decision, requiring forty-three page of text, concluded:

> There was a total of 1296.6 cubic yards of this trench excavation. We are of the opinion that the plaintiff is entitled to recover therefore the difference between the amount paid it at 69 cents a cubic yard and the amount due it at $4.00 a cubic yard. Plaintiff is entitled to recover on this item the sum of $4,291.75.[74]

There were increased costs that the U.S. Government could recover, such as Social Security payments and interest payments. This reduced the net amount to the contractor to only $1,927.59.

On a much happier note, an appreciative audience gathered on a high hill at the end of the Bolivar Dam ten miles north of New Philadelphia, Ohio, to celebrate the completion of the Muskingum Watershed Project. The ceremony was carried on nationwide radio and featured speeches by Major General Julian L. Schley, chief of the U.S. Army Corps of Engineers; John C. Page, U.S. Commissioner of Reclamation; John J. Jaster Jr.; and Dr. H. H. Bennett of the Soil Conservation Service; together with Ohio representatives

Senator Robert J. Bulkley and State Highway Director Robert N. Wilkin. It is interesting to note that the newspaper did not record the president of the MWCD speaking at this event, which seems to be a significant omission.

General Schley noted that the construction of the fourteen dams and reservoirs had required the relocation of 145 miles of highways, sixty miles of railway lines, and innumerable miles of pipes and wires. The occasion, he said, marked the completion of a twenty-five-year-old dream. Adding to Schley's comments, Commissioner Pace said ". . . the influence of the Muskingum Watershed Conservancy will be felt far beyond its limits. Its influence will be felt directly in a reduction of flood peaks down the Ohio River. . . . indirectly it will be felt from Maine to California. The Muskingum Valley is greatly advanced over most areas."[75]

On July 1, 1938, the operation of the dam was turned over to the MWCD. This grand gesture did not last long for, by August 22, 1939, the Corps of Engineers assumed operation of the dams and retained limited property around each of these structures. In transferring the property, it was agreed that the dam would be operated in strict conformance with the approved "official plan." This transfer was a result of the Flood Control Act of 1939. It eliminated the requirements of local funding on a matching basis for construction costs. As a result, the MWCD was reimbursed by $1.5 million and was relieved of both operating and maintenance cost of the dams. Thus, the Army Corps of Engineers could not control water releases primarily to benefit flood levels in the Ohio Valley. The system has been operated on this basis to the present day. At the beginning of the new year in 1940, the reservoirs at Beach City, Charles Mill, Dover, and Wills Creek were up to the conservation water levels. A year later, Dover was redesignated as a dry reservoir as a result of heavy industrial water pollution. With the completion of the watershed project, the Zanesville District Office was closed and Lieutenant Colonel Arthur transferred to the Huntington District as district engineer.[76] Effective October 1, 1940, Arthur left the post of Huntington District engineer to assume the position of head of the River and Harbor Board in Washington, D.C.

At the time of Arthur's transfer to Washington, the storm clouds of war had already enveloped Europe and threatened to involve the United States in yet another global conflict. With America on a war footing even before "the day of infamy" proclaimed by President Roosevelt, no large-scale public works were undertaken, and the New Deal faded away. In considering the historic significance of the Muskingum Project, one should consider that many currents in water-resource management find a confluence in the early Muskingum Navigation and later Watershed Flood Control Project. In *A Valley Renewed*, Hal Jenkins proclaims the success of the project:

> Time will come when the mental and moral forces which won in this valley the victory for flood control and water conservation will have as much weight in thoughtful minds as the site of those dams and reservoirs, even when all of the district's 14 improvements shall have been seen, one after the other.[77]

Muskingum Watershed Dams: Construction Dates

Reservoir/Dam Name	Starting Date	Completion Date
Atwood	12 August 1935	23 September 1937
Beach City	25 March 1935	13 August 1937
Bolivar	26 September 1935	29 September 1938
Charles Mill	11 January 1935	17 August 1936
Clendening	4 March 1935	1 November 1937
Dover	6 January 1935	13 November 1938
Leesville	13 June 1935	22 October 1937
Mohawk	1 May 1935	22 September 1937
Mohicanville	15 June 1935	6 August 1937
Piedmont	1 January 1935	22 May 1937
Pleasant Hill	5 September 1935	22 January 1938
Senecaville	10 May 1935	14 May 1937
Tappan	4 January 1935	24 October 1936
Wills Creek	23 February 1935	13 October 1937

In this great public work, the Corps of Engineers worked out the practical details of a changing role in water-resource management. From a narrow focus on providing navigation on numerous rivers to an expanded role in flood control; hydroelectric power generation; soil conservation; community and industrial water supply; in selected cases the means of irrigation and, of course, navigation, the Corps assumed a much broader mission. A series of floods provided the stimulus for the 1927 Flood Control Act and the 1936 Act, in a sense freeing the Corps of Engineers from their sole focus on navigation issues. Floods also laid to rest the indefensible but traditional "levees-only" policy of the Corps of Engineers and led to it becoming an active member in the Muskingum Watershed Project.

The application of newly established soil mechanics as primary considerations in both engineering design and supervision of construction led to a transformation of design and construction methods of embankment dams. The earlier empirical approach gave way to one based solidly on engineering analysis. The fourteen dams stand as a silent witness to the success of this application of soil mechanics. Linking the new technologies directly to the pioneering work of Professor Karl von Terzaghi, founder of the discipline of soil mechanics in civil engineering, was the director in charge of the soil mechanics lab, Robert R. Philippe. Philippe was one of Terzaghi's students.

Also impressive is the sophistication of the hydraulic design that determined the reservoir capacity, the conduit and gate design, and the safety features embodied in the spillways.

Early in the canal era, during the first half of the nineteenth century, the state of Ohio built its canal network, which included the Ohio and Erie Canal passing through part of the Muskingum watershed and connecting to the Muskingum slackwater system. The state's canal and slackwater developments were the first attempts to put water resources to the use of transportation and industry. The state of Ohio, an enlightened leader, passed legislation in the twentieth century establishing conservancy districts to control flooding and also promote a broader mission of natural-resources conservation. The first district, the Miami Conservancy, was followed by the Muskingum Watershed Conservation District. This later became a model of a cooperative venture between the conservancy and the Corps of Engineers as leading partners in the fourteen dams and reservoirs projects and included other state and federal agencies such as the federal Soil Conservation Service and the Ohio forestry and highway departments. This organization was touted nationally as being a model for other flood-control and conservation projects.

Even with an enlightened vision, it is difficult to imagine that such an extensive public work could have been built without the New Deal work-relief program. With advance planning already in hand, the Muskingum Watershed Conservancy District was a natural early candidate for Roosevelt's large-scale public-works projects. The Muskingum project is a particularly successful case of the numerous New Deal projects undertaken during the 1930s.

The subsequent development of recreational areas, improved agricultural practices, and afforestation are notable features of this district that set a historical precedent for such activities across the nation. The keys to the entire comprehensive development are the individual dams owned and operated by the Corps of Engineers. By any standards they are collectively eligible for a theme nomination to the National Register of Historic Places.

Summary of Historical Significance:

1. Together with the Tygart Dam and Bluestone Dam in West Virginia, the Muskingum dams completed in 1938 were some of the earliest New Deal public works; thus, they were a prototype of federal flood-control activities in the Ohio River basin. As the first of this type of project, they became planning, design, and construction models for similar dams built throughout the nation.

2. As one of the earliest examples of a New Deal public-works project, the Tappan Dam and Reservoir was the first federal flood-control dam constructed in the Ohio basin.

3. The entire watershed project of fourteen dams and reservoirs is a notable example of cost sharing among federal, state, and local governments to finance and administer flood-control activities. The watershed comprises nearly twenty percent of the land area of the state of Ohio.

4. During the New Deal, the grant for the Muskingum Watershed Project was the only one for local water management under a grant from the National Industrial Recovery Act.

5. From an engineering point of view, there were a number of innovative features in the watershed project. A soil-testing laboratory, together with the application of the new science of soil mechanics, allowed engineers to put the construction of rolled compacted embankment dams on a new scientific basis, which achieved the greatest safety and economy known throughout the Corps of Engineers.

6. An important innovation of considerable engineering value was the development of empirical methods for determining flood levels throughout the basin. By using the Muskingum coefficient method, it was possible to determine flood levels at various locations along the Muskingum River. Such

methods allowed a more accurate and economical design of conduits and spillways at the various reservoirs.

7. Various control devices were installed in the embankment dams such as piezometer tubes to determine the saturation line and methods to detect any movement that would occur on either the up- or downstream faces of the dam. The method also confirmed the theoretical analysis of the saturation line.

NOTES

1. Official Plan for the Muskingum Watershed Conservancy District, 8 Oct 1934, amended 15 April 1935 and revised 5 June 1935; RG 77, Entry 111 Bulkies, File 7245–46/1 (Muskingum River), NARA II, 1–143.

2. Official Plan, 1–143.

3. Hal Jenkins, *A Valley Renewed* (Kent, OH: Kent State University Press, 1976), 118.

4. "Experts Are Employed on Technical Details Conservancy Project" 5 Aug 1934, Schneider Scrapbook.

5. Analysis of Design, Beach City Dam, RG 77, Entry 111, File 3524/44-A, 14, 15, 18, NARA II.

6. Analysis of Design, Beach City Dam, 16.

7. Analysis of Design, Beach City Dam, 1–143.

8. Report on M.K.W. Model Study, Rolled Fill Section, U.S. Engineer Soil Laboratory, Zanesville, OH, April 1935; RG 77, Entry 111, Bulkies, File 7245, 2, NARA II.

9. Report on M.K.W. Model Study, 2.

10. Board of Engineers Meeting, 12 Dec 1934, as reported in Board of Engineers Meeting 28–31 Jan 1935, 6–8.

11. Board of Engineers Meeting, 12 Dec 1934, 6–8.

12. Board of Engineers Meeting, 28–31 Jan 1935, 1–8 + graphs.

13. Analysis of Design, Dover Dam, RG 77, Entry 111, File 3524–10/1 (Muskingum R. Dover Dam), 1, NARA II.

14. "Ohio Starts Dams to Control Floods," *New York Times*, 6 Jan 1935; "Zoarville, Town of 40 Houses, Must Move, Lake on Town Site," *Zanesville Times-Signal*, 24 Feb 1935; "Zoar Will Protest Sale Down River for Flood Project," 20 Aug 1935. All three articles from Schneider Scrapbook.

15. Senator Robert J. Bulkley, Washington, D.C., to Major General Edward M. Markham, Chief Engineer, Washington, D.C., 17 Jan 1935; RG 77, Entry 111, File 7249–36 (Muskingum R., O.), NARA II.

16. Acting Chief Engineer Brigadier General G. B. Pillsbury to Senator Robert J. Bulkley, 21 Jan 1935; RG 77, Entry 111, File 7249–36 (Muskingum R., O.), NARA II.

17. Acting Chief Engineer Brigadier General G. B. Pillsbury to Senator Robert J. Bulkley, 21 Jan 1935; RG 77, Entry 111, File 7249–36 (Muskingum R., O.), NARA II.

18. Jenkins, *A Valley Renewed*, 82.

19. E. T. Haines Abstract of Title (Ha-115T and Ha-116T) to the Muskingum Watershed Conservancy District, 1938–1947, Muskingum Watershed Conservancy District Archives, New Philadelphia, OH.

20. E. T. Haines Abstract of Title.

21. E. T. Haines Abstract of Title.

22. E. T. Haines Abstract of Title.

23. E. T. Haines Abstract of Title.

24. E. T. Haines Abstract of Title.

25. J. Norris Stroud Abstract of Title (Ha-164-T and Ha-168-T) to the Muskingum Watershed Conservancy District, May 1935, Muskingum Watershed Conservancy District, New Philadelphia, Ohio.

26. Inter-Office Memo to Bryce Browning from Ralph G. Wright, Muskingum Watershed Conservancy District, Land Department, Muskingum Watershed Conservancy District Archives, New Philadelphia, Ohio, 14 Sept 1949.

27. *George Henry v. Muskingum Watershed Conservancy District*, Harrison County (Ohio) Court of Common Pleas, entry No. 10451.

28. *Muskingum Watershed Conservancy District v. George Henry*, Harrison County (Ohio) Court of Common Pleas, entry No. 10714.

29. "Forces Set in Motion for Foundation for Greater Zanesville," 3 Feb 1935, Schneider Scrapbook.

30. "Senecaville Contract Award to Await Land Adjustment," 5 March 1935, Schneider Scrapbook.

31. "Purchase Three Key Farms for Dam Site at Senecaville," 17 March 1935; "Legislator and Browning to Aid Senecaville Dam," *Zanesville Times-Signal* 3 March 1935, Schneider Scrapbook.

32. Analysis of Design of Dover Dam, 1–14.

33. "Hundred Men Added to Watershed Gangs: Work Starts Soon on Dam at Senecaville," 28 April 1935, Schneider Scrapbook.

34. Jenkins, *A Valley Renewed*, 79, 193; "Sandy Creek Dam Is Retained in Watershed Plan," 8 June 1935, Schneider Scrapbook.

35. "Ten Dams Under Way; Thousand Men Busy," 9 June 1935, Schneider Scrapbook.

36. "Train Load of Soil Down Stream in Six Days of Rain," 26 May 1935; "Survey Completed Last Year Shows Erosion Greatest in Southeast Ohio," 26 May [1935], Schneider Scrapbook.

37. "Farming Subsoil Is Costly Both in Time and Money," 1 Sept 1935, Schneider Scrapbook.

38. "Employment on Dams at New Peak—2391 Men; Seek Bids to Remove Bridge Piers from River," 1 Sept 1935, Schneider Scrapbook.

39. "Application for Approval of Projects for Work Relief Muskingum Watershed Conservancy District New Philadelphia, Ohio," 28 May 1935, 1, Schneider Scrapbook.

40. Telegram to Harold Ickes from Gus Kasch, 17 Aug 1935; RG 77, Entry 111, File 7249–48/1 (Muskingum R., O.), NARA II; Lucius Clay, for the Chief of Engineers, to Mr. Gus Kasch, 4 Sept 1935; RG 77, Entry 111, File 7249–47 (Muskingum R., O.), NARA II.

41. Thomas P. Carroll to Gus Kasch, 26 Aug 1935; RG 77, Entry 111, File 7249–47/2 (Muskingum R., O.), NARA II.

42. Conference with Board of Consulting Engineers, 13–15 Sept 1935; RG 77, Entry 111, File 7245–61/1 (Muskingum R., O.), NARA II.

43. Conference with Board of Consulting Engineers, 19–20 Sept 1935; RG 77, Entry 111, File 7245–61/1 (Muskingum R., O.), NARA II.

44. "Wills Creek and Senecaville Projects Show Progress; High Speed at Mohawk," 6 Oct 1935, Schneider Scrapbook; Leland R. Johnson, *Men, Mountains and Rivers*, 101.

45. "Repeat Progress for First Year of Watershed Work," 4 Jan 1936, Schneider Scrapbook.

46. Conference of the Board of Consultants, 23–24 Jan 1936; RG 77, Entry 111, File 7245–69/1 (Muskingum R.), NARA II; "$3,500,000 More is Allotted Flood Plan," 21 Jan 1936, Schneider Scrapbook.

47. Major L. P. Worall, Finance Department, to Chief of Engineers, Washington D.C., 12 Feb 1936; RG 77, Entry 111, File 7245–68 (Muskingum R., O.), NARA II.

48. Major J. D. Arthur, District Engineer, to Division Engineer, Ohio River Division, 10 Feb 1936; RG 77, Entry 111, File 7249–53 (Muskingum R., O.), NARA II.

49. Harold Ickes, Secretary of the Interior & F.E.A. Administrator, to Major General Edward N. Markham, Chief of Engineers, 24 Aug 1936; RG 77, Entry 111, File 7245–70 (Muskingum R.), NARA II.

50. Ickes to Markham, 24 Aug 1936.

51. Secretary of War, Estimate of Cost of Examinations, etc. of Streams Where

Power Development Appears Feasible, H.R. Doc. 308, 69th Cong., 1st Sess.; Flood Control Act, 15 May 1928.

52. Colonel R. G. Powell, Division Engineer to Chief of Engineers, Washington, D.C., 10 Sept 1936; RG 77, Entry 111, File 7249–54 (Muskingum R., O.), NARA II.

53. Major R. G. Moses, Assistant Mississippi River Commission, to Colonel Ernest Graves, Mississippi River Commission, 9 Oct 1936; RG 77, Entry 111, File 7249 (Muskingum River, Ohio, NARA II.

54. Brigadier General H. B. Ferguson, President, Mississippi River Commission, to Mississippi River Commission, Vicksburg, 9 Oct 1936; RG 77, Entry 111, File 7249 (Muskingum R., O.), NARA II.

55. Report on Muskingum River, Ohio (Huntington, U.S. Engineer Office, 1 Dec 1934), 1–24.

56. Arnold, *Evolution of the 1936 Flood Control Act*, 59–96.

57. Johnson, *Men, Mountains and Rivers*, 167–176.

58. Colonel R. G. Powell, Division Engineer, to Chief of Engineers, Washington, D.C., 20 Jan 1937; RG 77, Entry 111, File 7245–87 (Muskingum R., O.), NARA II.

59. Conference with Board of Engineers, 28–31 Jan 1937; RG 77, Entry 111, File 7245–89/1 (Muskingum R.), NARA II.

60. Glen Falls Indemnity Co. by B030 & Ritchie to United States Engineer, Washington, D.C., 12 May 1937; RG 77, Entry 111, File 3524–35 (Muskingum R.-Clendening Dam), NARA II.

61. Three Indorsements by Chief of Engineers, Ohio River Division Engineer & District Engineer, Lieutenant Colonel Arthur's Endorsement Contains a Quote from the Specifications, RG 77, Entry 111, File 3524–35 (Muskingum R.-Clendening Dam), NARA II.

62. Glen Falls Indemnity Co. to United States Engineer, 12 May 1937; RG 77, Entry 111, 61. File 3524–35 (Muskingum R.-Clendening Dam), NARA II.

63. 3rd Indorsement by Lieutenant Colonel Arthur to Division Engineer, Glen Falls Indemnity Co., 12 May 1937; RG 77, Entry 111, File 3524–35 (Muskingum R.-Clendening Dam), NARA II.

64. Change order to: Chief of Engineers, Washington, D.C., from Colonel R. G. Powell Division Engineer, 5 Nov 1937; RG 77, Entry 111, File 3616–76 (Muskingum R.-Beach City Dam), NARA II.

65. William Eisenberg & Sons to Chief of Engineers, Washington, D.C., 12 Oct 1936; RG 77, Entry 111 Bulkies, File 3524–39 (Muskingum R.-Beach City Dam), and, Colonel J. D. Arthur, District Engineer to William Eisenberg & Sons, Inc. Exhibit "C."

66. Preliminary Change Order (No Change in Contract Cost, Contract Time Extended), Contract No. W968eng.-75, 14 Nov 1936; RG 77, Entry 111, File 3524–117/24 (Muskingum R.-Beach City Dam), NARA II.

67. Lieutenant Colonel Arthur, District Engineer, to Chief of Engineers, Washington, D.C. (through the Division Engineer, Ohio River Division), 1 March 1938; RG 77, Entry 111, File 3616–81 (Muskingum R.-Beach City Dam), NARA II. Lieutenant Colonel Arthur to Chief of Engineers, Washington, D.C. (through Division Engineer, Ohio River Division), Change Order No. 20, 4 Aug 1938; RG 77, Entry 111, File 3616–83 (Muskingum R.-Beach City Dam), NARA II.

68. J. D. Arthur Jr., 4th Indorsement 3524–108 (Muskingum R.-Beach City Dam), 19 Aug 1938; RG 77, Entry 111, File 3524–117/28, (Muskingum R.-Beach City Dam), NARA II.

69. William Eisenberg & Sons to Chief of Engineers, Washington, D.C., 17 Dec 1938; RG 77, Entry 111, File 3524–113 (Muskingum R.-Beach City Dam), NARA II.

70. *William Eisenberg & Sons, Inc., to the use of the Aetna Casualty & Surety Company, Plaintiff v. The United States*, Defendant Case No. 44772.

71. In the Court of Claims: *William Eisenberg & Sons v. The United States*–Case No. 44772, 27 Dec 1941.

72. Captain Walter Pinkus to District Engineer, Huntington, West Virginia, 31 Dec 1941, RG 77, 72. Entry 111, File 3524–129 (Muskingum R.-Beach City dam), NARA II.

73. Reported by James A. Hoyt, Cases Decided in the Court of Claims of the United States 1 Feb to 30 April 1948, Vol. CX, Washington, D.C. 1948.

74. Reported by James A. Hoyt, Cases Decided in the Court of Claims of the United States 1 Feb to 30 April 1948, Vol. CX, Washington, D.C. 1948.

75. "Flood Project is Dedicated," 18 July 1938, Schneider Scrapbook.

76. "Colonel Arthur Assigned to Washington, D.C." 17 Sept 1940; "Colonel Arthur will Handle Atlantic Army Bases," 28 Dec 1940, Schneider Scrapbook.

77. Jenkins, *A Valley Renewed*, 170.

APPENDIX

Formation of the Watershed

To understand the history of the Muskingum watershed in terms of early navigational improvements and subsequent flood-control structures, one must rely upon the geosciences. This includes both geology and geography of not only the watershed itself but, in a larger sense, a view of the upper reaches of the Ohio River valley of which the Muskingum watershed is a contributory area.

The geosciences provide an understanding of the dentrated drainage system of the Muskingum River and its tributaries. This watershed is the largest in Ohio, draining one-fifth of the entire state, and covers much of the eastern half. The total watershed area is 8,040 square miles.[1]

Besides information on the formation of a watershed, geographical studies provide essential design data for rainfall and associated flooding. Not only are data on the magnitude of flooding essential for the design of appropriate flood-control measures such as dams, levees, and flood walls, but frequency of flooding is an important parameter to consider in engineering design. For example, a benefit/cost study would consider the benefits obtained from flood-control measures against the damage sustained in a severe flood. If such a flood occurred every 500 years, rather than every 50 years, the cost might outweigh the benefits.[2]

Knowledge of rainfall and soil conditions is also essential for successful agricultural and afforestation projects throughout the watershed. The fourteen flood-control dams, with one exception, were earth-filled dams constructed of local materials. Even the concrete Dover Dam was designed knowing the foundation materials that support this concrete dam and spillway.

The drainage history of the Ohio River watershed contains several stages. The first is the Teays-Mahomet pre-glacial drainage systems. The second is characterized as the deep stage, followed by the present or post-glacial epoch.[3] In the earliest epoch, sedimentary rocks of the Ohio Valley were laid down in a shallow inland sea during the Paleological era; geologists label this the Pre-Cambrian basement. These sedimentary deposits caused the basement rock to subside, creating the Appalachian syncline. Being near sea level, the shallow waters fostered the development of extensive swamps and associated peat deposits. These deposits were later transformed into the Appalachian bituminous coalfields.

Geologists believe that approximately 200 million years ago the syncline was uplifted by tectonic action, which effectively ended the Paleozoic era.[4] In the Teays-Mahomet system, many of the rivers flowed toward the north and emptied into the Great Lakes drainage system. These included the Allegheny, the Monongahela, and the upper Ohio River.[5] Farther down the river, the drainage was controlled by a most unusual river system. Rising in North Carolina, the Teays River cut across the uplifted ridges of the Appalachian Mountain range, following the modern course of the New River through what is now West Virginia. It continued across Ohio and Indiana into Illinois, where it joined the Illinois River system at Mahomet and emptied into the inland sea.[6] The source of this ancient river lies close to the Atlantic Ocean but, in a capricious manner, the river flows northwest.

In comparatively recent geological time, the Ice Age, ca. 20,000 years ago, dammed the Teays-Mahomet River system and blocked the north-flowing rivers. The glacial boundary passed through the Muskingum watershed, forming the Ohio River and its tributaries as we know them today. This turned the Monongahela and the Allegheny away from the Great Lakes watershed to form the Ohio River at Pittsburgh. The only remnant of the extensive Teays-Mahomet drainage system is the New River. The remaining system was buried deeply under a considerable overburden of glacial outwash. Only recently have geologists determined the course of this primeval river.[7]

As a result of this glaciation, the Muskingum watershed is a dentrated system of many tributaries that flow into the Ohio River at Marietta.[8]

Muskingum Watershed Climate

The climate of the upper Ohio River, which includes the Muskingum watershed, is quite variable, with hot, humid summers and cold winters. This continental climate is subject to frequent changes caused by major storms moving east with associated high rainfall. Lying as it does in the middle latitudes of the continent, it is in the path of predominantly westerly winds, the exception being the influence of the Gulf of Mexico, which supplies great quantities of warm, moist air.[9]

A typical storm is produced by warm, moist Gulf air colliding with cooler air from the north and northwest. In terms of flooding, this type of storm is particularly damaging. Although of a fairly low intensity, the rain is of long duration with wide distribution over the entire watershed, producing very large total amounts of rainfall. The weather systems that produce heavy rainfall are associated with a high-pressure zone off the Atlantic coast, coupled with an equal high-pressure zone in the Mississippi Basin or even in Canada. These two zones establish a stationary front blocking moist Gulf air, producing sustained and heavy rainfall when the Gulf air collides with the front.[10]

Such a storm occurred in Sugar Creek in the Muskingum Basin in 1935. Two large high-pressure zones established a stationary front that precluded Gulf air from moving east. As it cooled, abundant rainfall resulted: 7.5 inches in twelve hours fell over a 1,500-square-mile area and five inches over 5,000 square miles, with nearly forty percent in three hours and two-thirds in six hours. The isohyetal lines clearly indicate the magnitude and distribution of the storm, while the runoff, which is critical to the development of flooding, is shown as rainfall plotted against time.[11] Both the devastating floods of 1913 and 1927 passed through the watershed causing widespread damage. These storms were instrumental in redefining the role of the Corps of Engineers regarding flood control.[12]

Over the years, "empirical formulae" for predicting storm frequency have been developed by several investigators based upon actual rainfall data. Amongst these are formulae by Meyers, Schafmager and Grant, Yarnell, and Bernard.[13] Thus, in designing flood-control measures such as reservoirs and levees, engineers have an empirical means of predicting the amount of rainfall. The critical factor in flooding is the amount of runoff from a given amount of precipitation.

A number of factors influence this amount. The area of the watershed affects the rate of runoff. For large watersheds, the flood crest moves slowly but can remain at flood stage for days or even weeks. On the contrary, for small, steep watersheds, the runoff discharges quickly into streams and subsides with equal rapidity. The shape and orientation of the watershed with regard to the axis of the storm will influence how the runoff is collected and discharged. For example, a long, narrow watershed typically will consist of a series of relatively short tributaries quickly discharging runoff into a series of flood crests—if the duration of the storm is short. If, however, the storm is of longer duration, the flood crest of the upper end of the watershed will move downstream and join the flood crest of the lower tributaries, and at that stage the major flood discharge is obtained. For a compact watershed characterized by fewer but longer tributaries, the runoff discharge for those various tributaries will reach the main stream at nearly the same time, creating a maximum flood event.

Another important factor in flood runoff is the slope of the watershed; the steeper the slope, the greater the percentage of runoff for a given intensity of rainfall. Not only will the runoff be greater because less water will have time to be absorbed in the ground, but, equally important, the floodwater will reach the collecting watercourse more quickly. Plainly, the topography of a watershed influences the frequency and intensity of flooding.

It is obvious that the characteristics of the soil and surface will control the percentage of runoff. A paved parking area will produce 100 percent runoff, whereas a sandy soil may absorb most of the precipitation and thus greatly reduce the percentage of runoff. Closely related is the nature of ground vegetation. By loosening the soil and retaining runoff flow, ground cover can mitigate the effects of rainfall in the short term. Once, however, the ground and vegetation have become saturated, the benefit decreases rapidly. With regard to the effect of forests on floods, there are opposing schools of thought. Some argue that deforestation increases runoff and the frequency of flooding. Those supporting this position are, not surprisingly, advocates of afforestation. The other position, however, states a belief that there has been no perceptible influence on stream flows for either afforested or cleared landscapes. For example, neither the well-known Captain Hiram Chittenden of the U.S. Army Corp of Engineers nor French engineers investigating the 1910 flood of the Seine River in France believed in the benefits of forest cover. The French claimed it had a marginal benefit of about three percent, whereas Chittenden believed the moisture held in the debris of the forest floor made matters worse compared to vegetative ground cover such as grass. Thus, during the development of a watershed conservation plan in connection with the Muskingum flood-control design, afforestation of areas in the watershed became an issue.

As in the case of empirical formulae and graphs for predicting the intensity and frequency of rainfall causing flooding, numerous formulae were developed to aid engineers in designing measures to control flooding by predicting quantities of maximum runoff. Amongst these are formula by Fanning, Murphy, Cooley, Kuichling, Bremner, Harman, Pettis, and Fuller.[14] In certain situations, a more rational method can be used. If one assumes a uniform intensity over the entire watershed continuing until all areas are contributing to the discharge, then the rate of discharge for the unit area will equal the rate of runoff, resulting in a simple formula.[15]

Seldom does one encounter situations that lend themselves to an ideal solution. There is some doubt whether the so-called rational method yields more reliable results than empirical models, since both intensity of rainfall and the runoff factors are estimated. Nevertheless, by the time the Muskingum watershed flood-control designs were undertaken in the late 1930s, analytical tools were available to design flood-control dams and flood walls throughout the watershed. Such formulae served not only as the basis of engineering designs, but also for calculating construction cost estimates.

Floods

Although rainfall and flood-control data are lacking for the early decades of the nineteenth century, the table in chapter 3, page 53 gives a list of floods that occurred in the Muskingum watershed area over a period of more than a century. Two of the floods are important in the history of man's attempt to control and utilize the water resources of the watershed. The first is the disastrous flood of 1913, and the second is the 1927 flood, which ensured the role of the Corps of Engineers in flood control. There were, of course, other serious floods, such as the great floods of 1936 and a number of inundations in the nineteenth century, that caused considerable damage.

NOTES

1. War Department, Corps of Engineers, *Report on the Muskingum River, Ohio, Covering Navigation, Flood Control, Power Development, and Irrigation* (Huntington, WV: United States Engineering Office, 10 Dec 1932), 3–4; RG 77, Entry 111, File 7249 Bulkies (Preliminary Examinations–Muskingum River), 3–4, NARA—Philadelphia.

2. George W. Pickels, *Drainage and Flood Control Engineering* (New York: McGraw-Hill Book Co., 1941), 47.

3. Secretary of War, *Ohio River*, H.R. Doc. 306, 74th Cong., 1st sess., 1936, 49.

4. *Final Environmental Impact Statement, Ohio River Navigation Project, Operation and Maintenance.*

5. Secretary of War, *Ohio River*, 49.

6. Wilton N. Melhorn and John P. Kempton, eds., *Geology and Hydrogeology of the Teays-Mahomet Bedrock Valley System*, Special Paper 258 (Boulder, CO: Geological Society of America, 1991), 19–30.

7. Melhorn and Kempton, *Geology and Hydrogeology*, 3–8.

8. J. C. Krolczyk, *Gazetteer of Ohio Streams* (Columbus, OH: State of Ohio, Department of Natural Resources, 1954), 41–45.

9. Secretary of War, *Ohio River*, 21.

10. *Final Environmental Impact Statement*, 45.

11. William G. Hott and Walter B. Langbein, *Floods* (Princeton, NJ: Princeton University Press, 1955), 45.

12. E. K. Morse and Harold A. Thomas, "Floods in the Upper Ohio River and Tributaries," *Proceedings, American Society of Civil Engineers 83*, no. 3 (March 1937), 495–535.

13. Pickles, *Drainage and Flood Control Engineering*, 47–55.

14. Pickles, *Drainage and Flood Control Engineering*, 56–86.

15. Pickles, *Drainage and Flood Control Engineering*, 75–78.

ABOUT THE AUTHOR

Emory L. Kemp is Professor Emeritus of History and of Civil Engineering at West Virginia University. He is an international expert in the history of technology and the restoration of structures. He worked for leading engineering consulting firms in England before receiving his Ph.D. in Theoretical and Applied Mechanics from the University of Illinois. He joined West Virginia University in 1962 to establish a graduate program in structural engineering. He founded WVU's program in the history of science and technology. Fostering the use of a material-culture approach for the study of the industrial past, he has researched and preserved historic industrial sites around the country and overseas and has advocated their public interpretation. Kemp is a founding member and past president of the Society for Industrial Archaeology and past president of the Public Works Historical Society. He is a distinguished member of the American Society of Civil Engineers and has received numerous other prestigious awards.